CAMBRIDGE LIBRARY COLLECTION

Books of enduring scholarly value

Earth Sciences

In the nineteenth century, geology emerged as a distinct academic discipline. It pointed the way towards the theory of evolution, as scientists including Gideon Mantell, Adam Sedgwick, Charles Lyell and Roderick Murchison began to use the evidence of minerals, rock formations and fossils to demonstrate that the earth was older by millions of years than the conventional, Bible-based wisdom had supposed. They argued convincingly that the climate, flora and fauna of the distant past could be deduced from geological evidence. Volcanic activity, the formation of mountains, and the action of glaciers and rivers, tides and ocean currents also became better understood. This series includes landmark publications by pioneers of the modern earth sciences, who advanced the scientific understanding of our planet and the processes by which it is constantly re-shaped.

Biographical Account of James Hutton

James Hutton (1726–1797) was an eminent Scottish scientist known chiefly for his work in geology. Educated at Edinburgh University, Hutton then travelled to Europe to study medicine before going into industry. He spent over a decade farming his family property in Scotland before returning to academic and commercial life. Hutton became an established geologist who also published on chemistry, meteorology and philosophy as an active member of the Edinburgh Royal Society. This volume, first published in 1805, is a detailed and affectionate chronicle of Hutton's life by his close friend, geologist and mathematician John Playfair. The author recounts Hutton's academic career, speculates on the motivation behind his foray into farming and includes a detailed discussion of his main geological theories. With little of Hutton's correspondence and papers surviving, this account by an intimate contemporary is the key resource for studying the life of an intriguing figure in scientific history.

Cambridge University Press has long been a pioneer in the reissuing of out-of-print titles from its own backlist, producing digital reprints of books that are still sought after by scholars and students but could not be reprinted economically using traditional technology. The Cambridge Library Collection extends this activity to a wider range of books which are still of importance to researchers and professionals, either for the source material they contain, or as landmarks in the history of their academic discipline.

Drawing from the world-renowned collections in the Cambridge University Library, and guided by the advice of experts in each subject area, Cambridge University Press is using state-of-the-art scanning machines in its own Printing House to capture the content of each book selected for inclusion. The files are processed to give a consistently clear, crisp image, and the books finished to the high quality standard for which the Press is recognised around the world. The latest print-on-demand technology ensures that the books will remain available indefinitely, and that orders for single or multiple copies can quickly be supplied.

The Cambridge Library Collection will bring back to life books of enduring scholarly value (including out-of-copyright works originally issued by other publishers) across a wide range of disciplines in the humanities and social sciences and in science and technology.

Biographical
Account of
James Hutton

JOHN PLAYFAIR

CAMBRIDGE
UNIVERSITY PRESS

CAMBRIDGE UNIVERSITY PRESS

Cambridge, New York, Melbourne, Madrid, Cape Town,
Singapore, São Paolo, Delhi, Tokyo, Mexico City

Published in the United States of America by Cambridge University Press, New York

www.cambridge.org
Information on this title: www.cambridge.org/9781108072502

© in this compilation Cambridge University Press 2011

This edition first published 1797
This digitally printed version 2011

ISBN 978-1-108-07250-2 Paperback

BIOGRAPHICAL ACCOUNT

OF

JAMES HUTTON, M.D. F.R.S.Ed.

BY

JOHN PLAYFAIR, F.R.S.Ed.

PROFESSOR OF MATHEMATICS IN THE UNIVERSITY OF EDINBURGH.

BIOGRAPHICAL ACCOUNT

OF THE LATE

Dr JAMES HUTTON.

DR JAMES HUTTON was the fon of Mr WILLIAM HUT-
TON, merchant in Edinburgh, and was born in that city
on the 3d of June 1726. His father, a man highly refpected
for his good fenfe and integrity, and who for fome years held
the office of City Treafurer, died while JAMES was very young.
The care of her fon's education devolved of courfe on Mrs
HUTTON, who appears to have been well qualified for difcharging
this double portion of parental duty. She refolved to beftow
on him a liberal education, and fent him firft to the High
School of Edinburgh, and afterwards to the Univerfity, where
he entered as a ftudent of humanity in November 1740.

OF the mafters under whom he ftudied there, MACLAURIN
was by far the moft eminent, and Dr HUTTON, though he had
cultivated the mathematical fciences lefs than any other, never
mentioned the lectures of that celebrated Profeffor but in terms
of high admiration.

HE ufed alfo to acknowledge his obligations to Profeffor STE-
VENSON's Prelections on Logic; not fo much, however, for having made him a logician as a chemift. The fact that gold is
diffolved in *aqua regia*, and that two acids which can each of

A   them

them fingly diffolve any of the bafer metals, muft unite their
ftrength before they can attack the moft precious, was mentioned by the Profeffor as an illuftration of fome general doctrine.
The inftinct of genius, if I may call it fo, enabled Mr HUTTON,
young as he then was, to feel, probably, rather than to under-
ftand, the importance of this phenomenon ; and as if, by the
original conftitution of his mind, a kind of elective attraction
had drawn him towards chemiftry, he became from that moment attached to it by a force that could never afterwards be
overcome. He made an immediate fearch for books that might
give him fome farther inftruction concerning the fact which he
had juft heard of ; but the only one he could procure, for a long
time, was HARRIS's *Lexicon Technicum,* the predeceffor of thofe
voluminous compilations which have fince contributed fo much
more to extend the furface, than to increafe the folidity of
fcience. It was from the imperfect fketch contained in that
dictionary, that he derived his firft knowledge of chemiftry, his
love for which never forfook him afterwards, and was in truth
the propenfity which decided the whole courfe and complexion
of his future life.

THOUGH his tafte and capacity for inftruction were fufficiently confpicuous during his courfe of academical ftudy, his friends
wifhed him rather to purfue bufinefs than fcience. This was a
meafure by no means congenial to his mind, yet he acquiefced
in it without difficulty.

ACCORDINGLY, in 1743 he was placed as an apprentice with
Mr GEORGE CHALMERS, writer to the Signet ; and fubjection to
the *routine* of a laborious employment, was now about to check
the ardour and reprefs the originality of a mind formed for different purfuits. But happily the force of genius cannot always
be controlled by the plans of a narrow and fhort-fighted pru-
dence. The young man's propenfity to ftudy continued, and
he was often found amufing himfelf and his fellow apprentices

with

with chemical experiments, when he fhould have been copying papers, or ftudying the forms of legal proceedings; fo that Mr CHALMERS foon perceived that the bufinefs of a writer was not that in which he was deftined to fucceed. With much good fenfe and kindnefs, therefore, he advifed him to think of fome employment better fuited to his turn of mind, and releafed him from the obligations which he had come under as his apprentice. In this he did an effential fervice to fcience, and to the young man himfelf. A man of talents may follow any profeffion with advantage; a man of genius will hardly fucceed but in that which nature has pointed out.

THE ftudy of medicine, as being the moft nearly allied to chemiftry, was that to which young HUTTON now refolved to dedicate his time. He began that ftudy under Dr GEORGE YOUNG, the father of the late Dr THOMAS YOUNG, and at the fame time attended the lectures in the Univerfity. This courfe of medical inftruction he followed from 1744 to 1747.

THOUGH a regular fchool of medicine had now been eftablifhed in the Univerfity of Edinburgh for feveral years, the fyftem of medical education was neither in reality, nor in the opinion of the world, fo complete as it has fince become. Some part of a phyfician's ftudies was ftill to be profecuted on the Continent; and accordingly, in the end of 1747, Mr HUTTON repaired to Paris, where he purfued with great ardour the ftudies of chemiftry and anatomy. After remaining in that metropolis nearly two years, he returned by the way of the Low Countries, and took the degree of Doctor of Medicine at Leyden in September 1749. His thefis is entitled, *De Sanguine et Circulatione in Microcofmo.*

ON his return to London about the end of that year, he began to think ferioufly of fettling in the world. His native city, to which his views of courfe were firft turned, afforded no very flattering profpect for his eftablifhment as a phyfician. The

A 2 bufinefs

busineſs there was in the hands of a few eminent practitioners who had been long eſtabliſhed; ſo that no opening was left for a young man whoſe merit was yet unknown, who had no powerful connections to aſſiſt him on his firſt outſet, and very little of that patient and circumſpect activity by which a man puſhes himſelf forward in the world.

THESE conſiderations ſeem to have made a very deep impreſſion on his mind, and he wrote on the ſubject of his future proſpects with conſiderable anxiety to his friends in Edinburgh.

ONE of theſe friends was Mr JAMES DAVIE, a young man nearly of his own age, with whom he had early contracted a very intimate friendſhip, that endured through the whole of his life, without interruption, to the mutual benefit of both. The turn which both of them had for chemical experiments formed their firſt connection, and cemented it afterwards. They had begun together to make experiments on the nature and production of ſal ammoniac. Theſe experiments had led to ſome valuable diſcoveries, and had been farther purſued by Mr DAVIE during Dr HUTTON's abſence. The reſult afforded a reaſonable expectation of eſtabliſhing a profitable manufacture of the ſalt juſt named from coal-ſoot.

THE project of this eſtabliſhment was communicated by Mr DAVIE to his friend, who was ſtill in London, and it appears to have leſſened his anxiety about ſettling as a phyſician, and probably was one of the main cauſes of his laying aſide all thoughts of that profeſſion. Perhaps, too, on a nearer view, he did not find that the practice of medicine would afford him that leiſure for purſuing chemical and other ſcientific objects, which he fancied it would do when he ſaw things at a greater diſtance. Whatever was the cauſe, it is certain that ſoon after his return to Edinburgh in ſummer 1750, he abandoned entirely his views of the practice of medicine, and reſolved to apply himſelf to agriculture.

THE

THE motives which determined him in the choice of the latter, cannot now be traced with certainty. He inherited from
his father a fmall property in Berwickfhire, and this might fug-
geft to him the bufinefs of hufbandry. But we ought rather, I
think, to look for the motives that influenced him, in the fimplicity of his character, and the moderation of his views, than in
external circumftances. To one who, in the maturity of under-
ftanding, has leifure to look round on the various employments
which exercife the fkill and induftry of man, if his mind is independent and unambitious, and if he has no facrifice to make to
vanity or avarice, the profeffion of a farmer may feem fairly entitled to a preference above all others. This was exactly the
cafe of Dr HUTTON, and he appears to have been confirmed in
his choice by the acquaintance which he made about that time
with Sir JOHN HALL of Dunglafs, a gentleman of the fame
county, a man of ingenuity and tafte for fcience, and alfo much
converfant with the management of country affairs.

As he was never difpofed to do any thing by halves, he determined to ftudy rural economy in the fchool which was then
reckoned the beft, and in the manner which is undoubtedly the
moft effectual. He went into Norfolk, and fixed his refidence for
fome time in that country, living in the houfe of a farmer, who
ferved both for his landlord and his inftructor. This he did in
1752; and many years afterwards I have often heard him mention, with great refpect, the name of JOHN DYBOLD, at whofe
houfe he had lived with much comfort, and whofe practical
leffons in hufbandry he highly valued. He appears, indeed, to
have enjoyed this fituation very much: the fimple and plain
character of the fociety with which he mingled, fuited well
with his own, and the peafants of Norfolk would find nothing
in the ftranger to fet them at a diftance from him, or to make
them treat him with referve. It was always true of Dr HUT-

TON,

TON, that to an ordinary man he appeared to be an ordinary man, poffeffing a little more fpirit and livelinefs, perhaps, than it is ufual to meet with. Thefe circumftances made his refidence in Norfolk greatly to his mind, and there was accordingly no period of his life to which he more frequently alluded, in converfation with his friends; often defcribing, with fingular vivacity, the rural fports and little adventures, which, in the intervals of labour, formed the amufement of their fociety.

WHILE his head-quarters were thus eftablifhed in Norfolk, he made many journeys on foot into different parts of England; and though the main object of thefe was to obtain information in agriculture, yet it was in the courfe of them that to amufe himfelf on the road, he firft began to ftudy mineralogy or geology. In a letter to Sir JOHN HALL, he fays that he was become very fond of ftudying the furface of the earth, and was looking with anxious curiofity into every pit, or ditch, or bed of a river that fell in his way; " and that if he did not always avoid the fate of THALES, his misfortune was certainly not owing to the fame caufe." This letter is from Yarmouth; it has no date, but it is plain from circumftances, that it muft have been written in 1753.

WHAT he learned in Norfolk made him defirous of vifiting Flanders, the country in Europe where good hufbandry is of the oldeft date. He accordingly fet out on a tour in that country, early in fpring 1754, and travelling from Rotterdam through Holland, Brabant, Flanders, and Picardy, he returned to England about the middle of fummer. He appears to have been highly delighted with the garden culture which he found to prevail in Holland and Flanders, but not fo as to undervalue what he had learnt in England. He fays in a letter to Sir JOHN HALL, written foon after his arrival in London, " Had I doubted of it before I fet out, I fhould have returned fully convinced that they are good hufbandmen in Norfolk."

THOUGH

Though his principal object in this excursion was to acquire information in the practice of husbandry, he appears to have bestowed a good deal of attention on the mineralogy of the countries through which he passed, and has taken notice in his *Theory of the Earth* of several of the observations which he made at that time.

About the end of the summer he returned to Scotland, and hesitated a while in the choice of a situation where he might best carry into effect his plans of agricultural improvement. At last he fixed on his own farm in Berwickshire, and accordingly set about bringing it into order with great vigour and effect. A ploughman whom he brought from Norfolk set the first example of good tillage which had been seen in that district, and Dr Hutton has the credit of being one of those who introduced the new husbandry into a country where it has since made more rapid advances than in any other part of Great Britain.

From this time till about the year 1768, he resided for the most part on his farm, visiting Edinburgh, however, occasionally. The tranquillity of rural life affords few materials for biographical description ; and an excursion to the North of Scotland, which he made in 1764, is one of the few incidents which mark an interval of fourteen years, passed mostly in the retirement of the country. He made this tour in company with Commissioner afterwards Sir George Clerk, a gentleman distinguished for his abilities and worth, with whom Dr Hutton had the happiness to live in habits of the most intimate friendship. They set out by the way of Crieff, Dalwhinnie, Fort Augustus, and Inverness ; from thence they proceeded through East-Ross into Caithness, and returned along the coast by Aberdeen to Edinburgh. In this journey Dr Hutton's chief object was mineralogy, or rather geology, which he was now studying with great attention.

For

For feveral years before this period, Dr HUTTON was concerned in the fal-ammoniac work, which had been actually eftablifhed on the foundation of the experiments already mentioned, but remained in Mr DAVIE's name, only, till 1765: at that time a copartnerfhip was regularly entered into, and the work carried on afterwards in the name of both.

HE now found that his farm was brought into the regular order which good hufbandry requires, and that as the management of it became more eafy, it grew lefs interefting. An occafion offering of letting it to advantage, he availed himfelf of it. About the year 1768 he left Berwickfhire entirely, and became refident in Edinburgh, giving his undivided attention from that time to fcientific purfuits.

AMONG other advantages which refulted to him from this change of refidence, we muft reckon that of being able to enjoy, with lefs interruption, the fociety of his literary friends, among whom were Dr BLACK, Mr RUSSEL, profeffor of Natural Philofophy, Profeffor ADAM FERGUSON, Sir GEORGE CLERK, already mentioned, his brother Mr CLERK of Elden, Dr JAMES LIND, now of Windfor, and feveral others. Employed in maturing his views, and ftudying nature with unwearied application, he now paffed his time moft ufefully and agreeably to himfelf, but in filence and obfcurity with refpect to the world. He was, perhaps, in the moft enviable fituation in which a man of fcience can be placed. He was in the midft of a literary fociety of men of the firft abilities, to all of whom he was peculiarly acceptable, as bringing along with him a vaft fund of information and originality, combined with that gayety and animation which fo rarely accompany the profounder attainments of fcience. Free from the interruption of profeffional avocations, he enjoyed the entire command of his own time, and had fufficient energy of mind to afford himfelf continual occupation.

A

A GOOD deal of his leifure was now employed in the profecu-tion of chemical experiments. In one of thefe experiments, which he has no where mentioned himfelf, but which I have heard of from Dr BLACK, he difcovered that mineral alkali is contained in zeolite. On boiling the gelatinous fubftance ob-tained from combining that foffil with muriatic acid, he found that, after evaporation, fea-falt was formed. Dr BLACK did not recollect exactly the date of this experiment, but from circum-ftances judged that it was earlier than 1772: It is, if I miftake not, the firft inftance of an alkali being difcovered in a ftony body. The experiments of M. KLAPROTH and Dr KENNEDY have confirmed this conclufion, and led to others of the fame kind.

IN 1774 he made a tour through part of England and Wales, of which, I find no memorandum whatever among his papers. I know, however, that at this time he vifited the falt-mines in Chefhire, and made the curious obfervation of the concentric circles marked on the roof of thefe mines, to which he has referred in his *Theory of the Earth,* as affording a proof that the falt rock was not formed from mere aqueous depofition. His friend Mr WATT of Birmingham accompanied him in his vifit to the mines.

IT was after returning to Birmingham from Chefhire, that he fet out on the tour into Wales. One of the objects of this tour, as I learnt from himfelf, was to difcover the origin of the hard gravel of granulated quartz, which is found in fuch vaft abun-dance in the foil about Birmingham, and indeed over a great tract of the central part of England. This gravel is fo unlike that which belongs to a country of fecondary formation, that it very much excited his curiofity; and his prefent journey was undertaken with a view to find out whether among the primi-tive mountains of Wales, there were any that might be fuppo-fed to have furnifhed the materials of it. In Wales, however,

B　　　　　　　　　　　　　he

he faw none that could, with any probability, be fuppofed to have done fo; and he was equally unfuccefsful in all the other parts he visited, till returning, at a fmall diftance from Birmingham, the place from whence he had fet out, he found a rock of the very kind which he had been in fearch of. It belongs to a body of ftrata apparently primary, which break out between Broomfgrove and Birmingham, and have all the characters of the indurated gravel in queftion. If, however, they have furnifhed the materials of that gravel, it feems probable that it has been through the medium of the red fand-ftone, which abounds in thofe countries *.

In 1777 Dr Hutton's firft publication was given to the world, viz. a fmall pamphlet, intituled, *Confiderations on the Nature, Quality, and Diftinctions of Coal and Culm.* This little work, an octavo pamphlet of 37 pages, was occafioned by a queftion that had arifen, Whether the fmall coal of Scotland is the fame with the culm of England? and, Whether of courfe, like the latter, it is entitled, when carried coaftwife, to an exemption from the duty on coal? Some of the fmall coal from the Frith of Forth, which had been carried to the northern counties for the purpofe of burning lime, had been confidered by the revenue officers as liable to the fame duty with other coal, while the proprietors contended that it ought only to pay the lighter duty levied on culm. This was warmly difputed; and after occupying for fome time the attention of the Board of Cuftoms in Scotland, was at laft brought before the Privy Council.

Dr Hutton's pamphlet was intended to fupply the information neceffary for forming a judgment on this queftion. It is very ingenious and fatisfactory, though perhaps, confidering the purpofe for which it was written, it is on too fcientific a plan,

and

---

* Illuftrations of the Huttonian Theory, p. 375.

and conducted too strictly according to the rules of philosophical analysis. It proves that culm is the small, or refuse, of the infusible, or stone coal, such as that of Scotland for the most part is; that the small of the fusible coal, by caking or uniting together, becomes equally serviceable with the large coal; whereas the small of the infusible, by running down like loose sand, cannot be made to burn in the ordinary way, and is useful but for few purposes, so that it has been properly exempted from the usual duty on coal. A criterion is also pointed out for determining when small coal is to be regarded as culm, and when it may be considered as coal;—if, when a handful of it is thrown into a red-hot shovel, the pieces burn without melting down or running together, it decidedly belongs to the former *.

In the conclusion, an exemption from duty was obtained for the small coal of Scotland, when carried coastwise, and this regulation was owing in a great degree to the satisfactory information contained in Dr Hutton's pamphlet. It was a step, also, toward the entire abolition of those injudicious duties which had been so long levied on coal, when carried by sea beyond a certain distance from its native place. This abolition happened several years after the period we are speaking of, much to the benefit of the country, and to the credit of the administration under which it took place.

As Dr Hutton always took a warm interest in whatever concerned the advancement of the arts, particularly in his native country, he entered with great zeal into the project of an internal navigation between the Friths of Forth and Clyde. The comparative merit of the different plans, according to which that work was to be executed, gave rise to a good deal of discussion, and even of controversy. In these debates Dr Hutton

B 2

took

* A few copies of the *Considerations on Culm* are still to be found in the shop of C. Elliot, Edinburgh.

took a fhare, and wrote feveral pieces, in which the grave and the ludicrous were both occafionally employed. None of thefe pieces have been publifhed; but the plan that was in the end adopted was that in favour of which they were written. It is unneceffary, however, to enter into the merits of a queftion which has long ceafed to intereft the public.

FROM the time of fixing his refidence in Edinburgh, Dr HUT-TON had been a member of the Philofophical Society, known to the world by the three volumes of phyfical and literary effays fo much and fo juftly efteemed *. In that fociety he read feveral papers; but it was during the time that elapfed between the publication of the laft of the volumes juft mentioned, and the incorporation of the Philofophical into the Royal Society of Edinburgh; which laft was eftablifhed by a royal charter in 1783. None of thefe papers have been publifhed, except one in the fecond volume of the *Tranfactions of the Royal Society*, " On certain Natural Appearances of the Ground on the Hill of Arthur's Seat."

THE inftitution of the Royal Society of Edinburgh had the good effect of calling forth from Dr HUTTON the firft fketch of a theory of the earth, the formation of which had been the great ob-ject of his life. From the date formerly mentioned, when he was yet a very young man, and making excurfions on foot through the different counties of England, till that which we are now arrived at, a period of about thirty years, he had never ceafed to ftudy the natural hiftory of the globe, with a view of afcertain-ing

* THE Philofophical Society was inftituted about the year 1739. The firft vo-lume of *Effays* was publifhed in 1754; the fecond in 1756; the third in 1771. From the year 1777 to 1782, the meetings of the Society were pretty regular, much owing to the zeal of Lord KAMES. Mr MACLAURIN may be regarded as the founder of this Society.

ing the changes that have taken place on its furface, and of dif-
covering the caufes by which they have been produced.

HE had become a fkilful mineralogift, and had examined the
great facts of geology with his own eyes, and with the moft
careful and fcrupulous obfervation. In the courfe of thefe ftudies he had brought together a confiderable collection of minerals peculiarly calculated to illuftrate the changes which foffil
bodies have undergone. He had alfo carefully perufed almoft
every book of travels from which any thing was to be learned
concerning the natural hiftory of the earth; and, in confequence
both of reading and obfervation, was eminently fkilled in phyfical geography.

IF to all this it be added, that Dr HUTTON was a good chemift, and poffeffed abilities excellently adapted to philofophical refearch, it will be acknowledged, that few men have entered with better preparation on the arduous tafk of inveftiga-
ting the true theory of the earth. Several years before the time
I am now fpeaking of, he had completed the great outline of
his fyftem, but had communicated it to very few; I believe to
none but his friends Dr BLACK and Mr CLERK of Elden.
Though fortified in his opinion by their agreement with him,
(and it was the agreement of men eminently qualified to judge),
yet he was in no hafte to publifh his theory; for he was
one of thofe who are much more delighted with the contempla-
tion of truth, than with the praife of having difcovered it. It
might therefore have been a long time before he had given any
thing on this fubject to the public, had not his zeal for fupport-
ing a recent inftitution which he thought of importance to the
progrefs of fcience in his own country induced him to come
forward, and to communicate to the Royal Society a concife account of his theory of the earth.

As

As I have treated of this theory in a feparate Effay, particularly deftined to the illuftration of it, I fhall here content myfelf with a very general outline.

I. The objeƈt of Dr HUTTON was not, like that of moft other theorifts, to explain the firft origin of things. He was too well fkilled in the rules of found philofophy for fuch an attempt; and he accordingly confined his fpeculations to thofe changes which terreftrial bodies have undergone fince the eftablifhment of the prefent order, in as far as diftinƈt marks of fuch changes are now to be difcovered.

WITH this view, the firft general faƈt which he has remarked is, that by far the greater part of the bodies which compofe the exterior cruft of our globe, bear the marks of being formed out of the materials of mineral or organized bodies, of more ancient date. The fpoils or the wreck of an older world are every where vifible in the prefent, and, though not found in every piece of rock, they are diffufed fo generally as to leave no doubt that the ftrata which now compofe our continents are all formed out of ftrata more ancient than themfelves.

II. The prefent rocks, with the exceptions of fuch as are not ftratified, having all exifted in the form of loofe materials colleƈted at the bottom of the fea, muft have been confolidated and converted into ftone by virtue of fome very powerful and general agent. The confolidating caufe which he points out is fubterraneous heat, and he has removed the objeƈtions to this hypothefis by the introduƈtion of a principle new and peculiar to himfelf. This principle is the compreffion which muft have prevailed in that region where the confolidation of mineral fubftances was accomplifhed. Under the weight of a fuperincumbent ocean, heat, however intenfe, might be unable to volatilize any part of thofe fubftances which, at the furface, and under the lighter preffure of our atmofphere, it can entirely confume. The fame preffure, by forcing thofe fubftances to remain united,

ted, which at the furface are eafily feparated, might occafion the fufion of fome bodies which in our fires are only calcined. Hence the objections that are fo ftrong and unanfwerable, when oppofed to the theory of volcanic fire, as ufually laid down, have no force at all againft Dr HUTTON's theory; and hence we are to confider this theory as hardly lefs diftinguifhed from the hypothefis of the Vulcanifts, in the ufual fenfe of that appellation, than it is from that of the Neptunifts, or the difciples of WER-NER.

III. THE third general fact on which this theory is founded, is, that the ftratified rocks, inftead of being either horizontal, or nearly fo, as they no doubt were originally, are now found poffeffing all degrees of elevation, and fome of them even perpendicular to the horizon; to which we muft add, that thofe ftrata which were once at the bottom of the fea are now raifed up, many of them, feveral thoufand feet above its furface. From this, as well as from the inflexions, the breaking and feparation of the ftrata, it is inferred, that they have been raifed up by the action of fome expanfive force placed under them. This force, which has burft in pieces the folid pavement on which the ocean refts, and has raifed up rocks from the bottom of the fea, into mountains 15,000 feet above its furface, exceeds any which we fee actually exerted, but feems to come nearer to the caufe of the volcano or the earthquake than to any other, of which the effects are directly obferved. The immenfe difturbance, therefore, of the ftrata, is in this theory afcribed to heat acting with an expanfive power, and elevating thofe rocks which it had before confolidated.

IV. AMONG the marks of difturbance in which the mineral kingdom abounds, thofe great breaches among rocks, which are filled with materials different from the rock on either fide, are among the moft confpicuous. Thefe are the veins, and comprehend, not only the metallic veins, but alfo thofe of whin-

ftone,

ftone, of porphyry, and of granite, all of them fubftances more
or lefs cryftallized, and none of them containing the remains of
organized bodies. Thefe are of pofterior formation to the ftrata
which they interfect, and in general alfo they carry with them
the marks of the violence with which they have come into their
place, and of the difturbance which they have produced on the
rocks already formed. The materials of all thefe veins Dr Hutton concludes to have been melted by fubterraneous heat, and,
while in fufion, injected among the fiffures and openings of
rocks already formed, but thus difturbed, and moved from their
original place.

This conclufion he extends to all the maffes of whinftone,
porphyry, and granite, which are interpofed among ftrata, or
raifed up in pyramids, as they often appear to be, through the
midft of them. Thus, in the fufion and injection of the un-
ftratified rocks, we have the third and laft of the great opera-
tions which fubterraneous heat has performed on mineral fubftances.

V. From this Dr Hutton proceeds to confider the changes
to which mineral bodies are fubject when raifed into the atmofphere. Here he finds, without any exception, that they are all
going to decay; that from the fhore of the fea to the top of the
mountain, from the fofteft clay to the hardeft quartz, all are
wafting and undergoing a feparation of their parts. The bodies thus refolved into their elements, whether chemical or mechanical, are carried down by the rivers to the fea, and are there
depofited. Nothing is exempted from this general law: among
the higheft mountains and the hardeft rocks, its effects are moft
clearly difcerned; and it is on the objects which appear the moft
durable and fixed, that the characters of revolution are moft
deeply imprinted.

On comparing the firft and the laft of the propofitions juft
enumerated, it is impoffible not to perceive that they are two

 fteps

steps of the same progreſſion, and that mineral ſubſtances are alternately diſſolved and renewed. Theſe viciſſitudes may have been often repeated; and there are not wanting remains among mineral bodies, that lead us back to continents from which the preſent are the third in ſucceſſion. Here, then, we have a ſeries of great natural revolutions in the condition of the earth's ſurface, of which, as the author of this theory has remarked, we neither ſee the beginning nor the end; and this circumſtance accords well with what is known concerning other parts of the economy of the world. In the continuation of the different ſpecies of animals and vegetables that inhabit the earth, we diſcern neither a beginning nor an end; and in the planetary motions, where geometry has carried the eye ſo far both into the future and the paſt, we diſcover no mark either of the commencement or termination of the preſent order. It is unreaſonable, indeed, to ſuppoſe that ſuch marks ſhould any where exiſt. The Author of nature has not given laws to the univerſe, which, like the inſtitutions of men, carry in themſelves the elements of their own deſtruction; he has not permitted in his works any ſymptom of infancy or of old age, or any ſign by which we may eſtimate either their future or their paſt duration. He may put an end, as he no doubt gave a beginning, to the preſent ſyſtem, at ſome determinate period of time; but we may reſt aſſured, that this great cataſtrophe will not be brought about by the laws now exiſting, and that it is not indicated by any thing which we perceive.

IT would be deſirable to trace the progreſs of an author's mind in the formation of a ſyſtem where ſo many new and enlarged views of nature occur, and where ſo much originality is diſplayed. On this ſubject, however, Dr HUTTON's papers do not afford ſo much information as might be wiſhed for, though ſomething may be learnt from a few ſketches of an Eſſay on the *Natural Hiſtory of the Earth*, evidently written at a very early period, and intended, it would ſeem, for parts of an extenſive

C　　　　　　　　　work,

work, of which, as often happens with the firſt attempts to ge-
neralize, the plan was never executed, and may never have been
accurately digeſted.

FROM theſe ſketches it appears that the firſt of the propoſitions
juſt enumerated, viz. that a vaſt proportion of the preſent rocks
is compoſed of materials afforded by the deſtruction of bodies,
animal, vegetable, and mineral, of more ancient formation, was
the firſt concluſion that he drew from his obſervations.

THE ſecond ſeems to have been, that all the preſent rocks
are without exception going to decay, and their materials de-
ſcending into the ocean. Theſe two propoſitions, which are the
extreme points, as it were, of his ſyſtem, appear, as to the order
in which they became known, to have preceded all the reſt.
They were neither of them, even at that time, entirely new propoſitions, though, in the conduct of the inveſtigation, and in the
uſe made of them, a great deal of originality was diſplayed.
The compariſon of them naturally ſuggeſted to a mind not
fettered by prejudice, nor ſwayed by authority, that they are
two ſteps of the ſame progreſſion ; and that, as the preſent continents are compoſed from the waſte of more ancient land, ſo,
from the deſtruction of them, future continents may be deſtined
to ariſe. Dr HUTTON accordingly, in the notes to which I al-
lude, inſiſts much on the perfect agreement of the ſtructure of
the beds of grit or ſandſtone, with that of the banks of uncon-
ſolidated ſand now formed on our ſhores, and ſhews that theſe
bodies differ from one another in nothing but their compact-
neſs and induration.

IN generalizing theſe appearances, he proceeded a ſtep farther,
conſidering this ſucceſſion of continents as not confined to one
or two examples, but as indefinitely extended, and the conſe-
quence of laws perpetually acting. Thus he arrived at the new
and ſublime concluſion, which repreſents nature as having pro-
vided for a conſtant ſucceſſion of land on the ſurface of the earth,

according

according to a plan having no natural termination, but calculated to endure as long as thofe beneficent purpofes, for which the whole is deftined, fhall continue to exift.

THIS conclufion, however, was but a fuggeftion, till the mechanifm was inquired into by which this grand renovation may be brought about, or by which loofe materials can be converted into ftone, and elevated into land. This led to an inveftigation of the mineralizing principle, or the caufe of the confolidation of mineral bodies: And Dr HUTTON appears accordingly, with great impartiality, and with no phyfical hypothefis whatever in his mind, to have begun with inquiring into the nature of the fluidity which fo many mineral fubftances feem to have poffeffed previous to the acquifition of their prefent form. After a long and minute examination, he came to the conclufion, That the fluidity of thefe fubftances has been what he terms SIMPLE, that is to fay, not fuch as is produced by combination with a folvent. The two general facts from which this conclufion follows, are, firft, that no folvent is capable of holding in folution all mineral fubftances, nor even all fuch varieties of them as are often united in the fame fpecimen; and, fecondly, that in the bodies compofed of fragments of other bodies, the confolidation is fo complete that no room is left for a folvent to have ever occupied. The fubftance, therefore, which was the caufe of the fluidity of mineral bodies, and prepared them for confolidation, muft have been one that could act on them all, which occupied no fpace within them, and could find its way through them, whatever was the degree of their compactnefs and induration. Heat is the only fubftance which has thefe properties; and is the only one, therefore, which, without manifeft contradiction, can be affigned as the caufe of mineral confolidation.

MANY difficulties, however, were ftill to be removed before this hypothefis was rendered completely fatisfactory; but in what order Dr HUTTON proceeded to remove them, the notes above

mentioned

mentioned do not enable me to ſtate. We may neverthelefs conjecture, with confiderable probability, what the ſtep was which immediately followed.

IT muſt have occurred to him, as an objection to the confolidation of minerals by ſubterraneous heat, that many ſubſtances are found in the bowels of the earth in a ſtate altogether unlike that into which they are brought by the action of our fires at the ſurface. Coal, for inſtance, by expofure to fire, has its parts diſſipated ; the aſhes which remain behind are a ſubſtance quite different from the coal itſelf ; and hence it would ſeem that this foſſil can never before have been ſubjected to the action of fire. But is it certain, (we may ſuppofe Dr HUTTON to have ſaid to himſelf), if the heat had been applied to the coal in the interior of the earth, at the bottom of the ſea, for example, that the ſame diſſipation of the parts would have taken place? Would not the greater compreſſion that muſt prevail in that region have prevented the diſſipation, at leaſt till a more intenſe heat was applied? And if the diſſipation was prevented, might not the maſs, after cooling, be very different from any thing that can be obtained by burning at the ſurface of the earth? It is plain that there is no reafon whatever for anſwering theſe queſtions in the negative. And, on the contrary, if the analogy of nature is confulted, if the fact of water requiring more heat to make it boil when it is more compreſſed, or the experiments with PAPIN's digeſter, be confidered, it will appear that the anſwer muſt be in the affirmative. Nay, it could not but ſeem reafonable to proceed a ſtep farther, and, as the mixture of ſubſtances is known in ſo many inſtances to promote their fuſibility, to ſuppofe that when the volatile parts of bodies were reſtrained, the whole maſs might be reduced into fuſion by heat, though, when theſe ſame parts were driven off, the reſiduum might be altogether infuſible. Thus coal, when the charcoal and bitumen are forced to remain in union, may very well be a fuſible ſubſtance,

though,

though, when the latter is permitted to efcape, the former becomes one of the moft refractory of all bodies.

In this way, and probably from this very inftance, the
effects of compreffion may have fuggefted themfelves to Dr
HUTTON. He would foon perceive that the fame principle
could be very generally applied, and that it afforded the folution of a difficulty concerning limeftone, fimilar to that which
has been juft ftated with refpect to coal. Limeftone is not found
in the bowels of the earth having the caufticity which it ac-
quires by the action of fire, and hence one might conclude that
it had never been expofed to the action of that element. But
the experiments of Dr BLACK, before his friend was engaged in
this geological inveftigation *, had proved that the caufticity
of lime depends on the expulfion of the aëriform fluid, fince
diftinguifhed by the name of carbonic gas, which compofes no
lefs than two-fifths of the whole. This great difcovery, which
has extended its influence fo widely over the fcience of chemiftry, alfo led to important confequences in geology; and Dr
HUTTON inferred from it, that ftrong compreffion might prevent the caufticity of lime, by confining the carbonic gas, even
when great heat was applied, and that, as has been fuppofed of
coal, the whole may have been melted in the interior of the
earth, fo as on cooling to acquire that cryftallized or fparry
ftructure which the carbonate of lime fo frequently poffeffes †.

It

* Dr BLACK's paper on magnefia, which contained this difcovery, was communicated to the Philofophical Society of Edinburgh in June 1755, and was publifhed in the fecond volume of their Effays, in the year following. Dr HUTTON had
at this time only begun his geological refearches. It was not, I imagine, till after
the year 1760 that they came to take the form of a theory.

† In the view here prefented of the principle of compreffion, as employed in
the Huttonian Theory, it is confidered as a hypothefis, conformable to analogy,
affumed for the purpofe of explaining certain phenomena in the natural hiftory of
the

IT is unneceſſary to carry our conjectures concerning the train of Dr HUTTON's diſcoveries to a greater length ; the development of the principles now enumerated, and the compariſon of the reſults with the facts obſerved in the natural hiſtory of minerals, led to thoſe diſcoveries, by a road that will be eaſily traced by thoſe who ſtudy his theory with attention.

IT might have been expected, when a work of ſo much originality as this Theory of the Earth, was given to the world, a theory which profeſſed to be the reſult of ſuch an ample and

accurate

the earth.   It reſts, therefore, as to its evidence, partly on its conformity to analogy, and partly on the explanation which it affords of the phenomena alluded to. In ſuppoſing that it derives probability from the laſt-mentioned ſource, we are far from aſſuming any thing unprecedented in ſound philoſophy.   A principle is often admitted in phyſics, merely becauſe it explains a great number of appearances ; and the theory of GRAVITATION itſelf reſts on no other foundation.

THE degree of this evidence will perhaps be differently appreciated, according to a man's habits of thinking, or the claſs of ſtudies in which he has been chiefly engaged.   To Dr HUTTON himſelf it appeared very ſtrong ; for he conſidered the fact of the liquefaction of mineral ſubſtances by heat as ſo completely eſtabliſhed, that it affords a full proof of the fuſibility of thoſe ſubſtances having been increaſed by the compreſſion which they endured in the bowels of the earth.   In his view of the matter, no other proof ſeemed neceſſary, and he did not appear to think that the direct teſtimony of experiment, could it have been obtained, would have added much to the credibility of this part of his ſyſtem.

FOR my part, I will acknowledge that the matter appears to me in a light ſomewhat different, and that though the arguments juſt mentioned are ſufficient to produce a very ſtrong conviction, it is a conviction that would be ſtrengthened by an agreement with the reſults even of ſuch experiments as it is within our reach to make.   It ſeems to me, that it is with this principle in geology, much as it is with the parallax of the earth's orbit in aſtronomy ; the diſcovery of which, though not neceſſary to prove the truth of the COPERNICAN SYSTEM, would be a moſt pleaſing and beautiful addition to the evidence by which it is ſupported.   So, in the Huttonian geology, though the effects aſcribed to compreſſion, are fairly deducible from the phenomena of the mineral kingdom itſelf, compared with certain analogies which ſcience has eſtabliſhed, yet the teſtimony of direct experiment would make the evidence complete, and would leave nothing that incredulity itſelf could poſſibly deſiderate.

accurate induction, and which opened up fo many views, interefting not to mineralogy alone, but to philofophy in general, that it
would have produced a fudden and vifible effect, and that men
of fcience would have been every where eager to decide concerning its real value. Yet the truth is, that it drew their attention very flowly, fo that feveral years elapfed before any one
fhewed himfelf publicly concerned about it, either as an ene-
my or a friend.

SEVERAL caufes probably contributed to produce this indifference. The world was tired out with unfuccefsful attempts
to form geological theories, by men often but ill informed of
the phenomena which they propofed to explain, and who proceeded alfo on the fuppofition that they could give an account
of the *origin* of things, or the firft eftablifhment of that fyftem
which is now the order of nature. Men who guided their inquiries by a principle fo inconfiftent with the limits of the hu-
man faculties, could never bring their fpeculations to a fatisfactory conclufion, and the world readily enough perceived their
failure, without taking the trouble to inquire into the caufe
of it.

TRUTH, however, forces me to add, that other reafons certainly contributed not a little to prevent Dr HUTTON's theory
from making a due impreffion on the world. It was propofed
too briefly, and with too little detail of facts, for a fyftem which
involved fo much that was new, and oppofite to the opinions
generally received. The defcriptions which it contains of the
phenomena of geology, fuppofe in the reader too great a knowledge of the things defcribed. The reafoning is fometimes em-
barraffed by the care taken to render it ftrictly logical; and the
tranfitions, from the author's peculiar notions of arrangement,
are often unexpected and abrupt. Thefe defects run more or
lefs through all Dr HUTTON's writings, and produce a degree of
obfcurity aftonifhing to thofe who knew him, and who heard

him·

him every day converfe with no lefs clearnefs and precifion; than animation and force. From whatever caufes the want of perfpicuity in his writings proceeded, perplexity of thought was not among the number ; and the confufion of his ideas can neither be urged as an apology for himfelf, nor as a confolation to his readers.

ANOTHER paper from his pen, a *Theory of Rain*, appeared alfo in the firft volume of the *Edinburgh Tranfactions*. He had long ftudied meteorology with great attention ; and this communication contains one of the few fpeculations in that branch of knowledge entitled to the name of *theory*.

DR HUTTON begins with fuppofing that the quantity of humidity, which air is capable of diffolving, increafes with its temperature. Now, this increafe muft either be in the fame ratio with the increafe of heat, in a lefs ratio, or in a greater : in other words, for equal increments of heat, the increments of humidity muft either conftitute a feries of which all the terms are equal to one another, or a feries in which the terms continually decreafe, or one in which they continually increafe *. If either of the two firft laws was that which took place in nature, a mixture of two portions of air, though each contained as much humidity as it was capable of diffolving, would never produce a condenfation of that humidity. According to the

firft

* To fpeak ftrictly, the law which connects the increments of humidity in the air with the increments of temperature, is not confined to any one of the three fuppofitions here made, but may involve them all. The humidity diffolved may be proportional to fome *function* of the heat, that varies in fome places fafter, and in others flower, than in the fimple ratio of the heat itfelf. Neverthelefs, for that extent to which obfervation reaches, the reafoning of DR HUTTON is quite fufficient to prove that it varies fafter ; or, in other words, that if a curve be fuppofed, of which the abfciffæ reprefent the temperature, and the ordinates the humidity, this curve, though it may in the courfe of its indefinite extent be in fome places confave and in others convex toward the axis, is wholly convex in all that part with which our obfervations are concerned.

firſt law, the temperature, the humidity, and the power of containing humidity, in the mixture, being all arithmetical means between the ſame quantities, as they exiſted previouſly to the mixture, the temperature produced would be exactly that which was required by the humidity to preſerve it in its inviſible form. If the ſecond law took place, the moiſture actually contained in the mixture would be leſs than the temperature was capable of ſupporting; ſo that inſtead of a condenſation of humidity, the air would become drier than before.

IF, on the other hand, the third law be that which takes place, after the mixture of two portions of air of different temperatures, the humidity will be greater than the temperature is able to maintain, and therefore a condenſation of it will follow. Now, the experience of every day proves, that the mixture of two portions of humid air of unequal temperatures, does indeed produce a condenſation of moiſture, and therefore we are authoriſed to conclude that the laſt-mentioned law is that which actually prevails *.

IT is obvious that this principle affords an explanation of the formation of clouds in the atmoſphere, and that currents of air,

D   or

---

* IT has been ſuppoſed that the chemical ſolution of humidity in air is neceſſarily implied in this theory of rain. The truth is, that the air is here conſidered only as the vehicle of the vapour, and that the tranſparent ſtate of the latter is ſuppoſed to depend on the temperature, or the quantity of heat; but whether that heat act on the vapour ſolely and directly, or indirectly, by increaſing the power of the air to retain it in ſolution, is, with reſpect to this theory, altogether immaterial.

DR HUTTON has indeed uſed the common language concerning the ſolution of humidity in air; but the ſuppoſition of ſuch ſolution is not eſſential to his theory. He ſeemed, indeed, to entertain doubts about the reality of that operation, founded on the circumſtance of evaporation taking place *in vacuo*. Experiments made by M. DALTON ſince the death of DR HUTTON, ſhew that there is great reaſon for ſuppoſing that the air has no chemical action whatever on the aqueous vapour contained in it. *Mancheſter Memoirs*, vol. v. p. 538.

or winds, of different temperatures, when they meet, muft produce fuch mixtures as have been defcribed, and give rife confequently to the condenfation of aqueous vapour. When the fupply of the humid air, entering into the mixture, is continued, the quantity of cloud formed will continually increafe, and the fmall globules of condenfed moifture, uniting into drops, muft defcend in rain.

BUT though we are thus in poffeffion of a principle by which rain may be certainly produced, yet whether it be the only one by which rain is produced may require fome farther inveftigation. Dr HUTTON accordingly, in order to determine this point, has entered into a very ample detail concerning the rain under different climates, and in different regions of the earth. The refult is, that the quantity of rain is, as nearly as can be eftimated, every where proportional to the humidity contained in the air, and the caufes which promote the mixture of different portions of air, in the upper regions of the atmofphere. Between the tropics, for inftance, the dry feafon is that in which the uniform current of the trade-wind meets with no obftruction in its circuit round the globe; and the rainy feafon happens when the fun approaches to the zenith, and when the fteadinefs of the trade-wind either yields to irregular variations, or to the ftated changes of the monfoons.

THUS too, (to mention another extreme cafe), in certain countries diftant from the fea, having little inequality of furface, and expofed to great heat, no rain whatever falls, and the fands of the defert are condemned to perpetual fterility. Even there, however, where a mountainous tract occurs, the mixture of different portions of air produces a depofition of humidity; perennial fprings are found; and the fertile vales of Fezzan or Palmyra are exempted from the defolation of the furrounding wildernefs.

THIS ingenious theory attracted immediate attention, and was valued for affording a diftinct notion of the manner in

which

which cold acts in caufing a precipitation of humidity. It met, however, from M. DE LUC with a very vigorous and determined oppofition; Dr HUTTON defended it with fome warmth, and the controverfy was carried on with more fharpnefs, on both fides, than a theory in meteorology might have been expected to call forth. For this Dr HUTTON had leaft apology, if greateft indulgence, on the fcore of temper, is due to the combatant who has the worft argument. The merits of the queftion cannot be confidered here: It is fufficient to remark, that they came ultimately to reft on a fingle point, Whether the refrigeration of air is carried on by the mixture of the cold and the hot air, or by the paffage of the heat itfelf, without fuch mixture, from one portion of air to another. If the former holds, Dr HUTTON's theory is eftablifhed; if the latter be true, M. DE LUC's objections may at leaft merit examination.

Now, it is certain, that if not the only, yet almoft the only, communication of heat through fluids, is produced by the mixture of one part of the fluid with another. The ftatical principle by which heat is thus propagated, was firft, I believe, accurately explained by Dr BLACK, and fince his time has been farther illuftrated by the experiments of Count RUMFORD. Thefe laft have led their ingenious author to conclude that heat has no tendency to pafs through fluids, otherwife than by the mixture of the parts of different temperature. The accuracy of this conclufion, in its full extent, may reafonably be queftioned; but this much of it is undoubtedly true, that when the particles of a body are at liberty to move freely among themfelves, the direct communication of heat, compared with the ftatical, is evanefcent, and may be regarded as a mere infinitefimal. M. DE LUC's objections are therefore of no weight.

THE *Theory of Rain* was republifhed by Dr HUTTON in his *Phyfical Differtations* feveral years afterwards, together with his anfwers to M. DE LUC, and feveral other meteorological tracts,

which

which contain many excellent examples of generalization, in a branch of natural hiſtory where it is more eaſy to accumulate facts, and more difficult to aſcertain principles, than in any other *.

AFTER

---

* It may be proper to mention here ſome uſeful obſervations in meteorology which Dr Hutton made, but of which he has given no account in any of his publications.

He was, I believe, the firſt who thought of aſcertaining the medium temperature of any climate by the temperature of the ſprings. With this view he made a great number of obſervations in different parts of Great Britain, and found, by a ſingular enough coincidence between two arbitrary meaſures, quite independent of one another, that the temperature of ſprings, along the eaſt coaſt of this iſland, varies nearly at the rate of a degree of Fahrenheit's thermometer for a degree of latitude. This rate of change, though it cannot be general over the whole earth, is probably not far from the truth for all the northern part of the temperate zone.

For eſtimating the effect which height above the level of the ſea has in diminiſhing the temperature, he alſo made a ſeries of obſervations at a very early period. By theſe obſervations he found that the difference between the ſtate of the thermometer in two places of a given difference of level, and not very diſtant, in a horizontal direction, is a conſtant quantity, or one which remains at all ſeaſons nearly the ſame, and is about 1° for 230 feet of perpendicular height.

I muſt, however, obſerve, that on verifying theſe obſervations, I have found the rate of the decreaſe of temperature a little ſlower than this, and very nearly a degree for 250 feet. This ſeems to hold for a conſiderable height above the earth's ſurface, and will be found to come pretty near the truth, to the height of five or ſix thouſand feet. It is not however probable that the diminution of the temperature is exactly proportional to the increaſe of elevation ; and it would ſeem that at heights greater than the preceding, the deviation becomes ſenſible ; the differences of heat varying in a leſs ratio than the differences of elevation.

In explaining this diminution of temperature as we aſcend in the atmoſphere, Dr Hutton was much more fortunate than any other of the philoſophers who have conſidered the ſame ſubject. It is well known that the condenſation of air converts part of the latent into ſenſible heat, and that the rarefaction of air converts part of the ſenſible into latent heat. This is evident from the experiment of the air-gun, and from many others. If, therefore, we ſuppoſe a given quantity of air to be ſuddenly tranſpoſed from the ſurface to any height above it, the air will expand on account of the diminution of preſſure, and a part of its heat becoming

latent,

AFTER the period of the two publications juft mentioned, Dr HUTTON made feveral excurfions into different parts of Scotland, with a view of comparing certain refults of his theory of the earth with actual obfervation. His account of granite, viz. that it is a fubftance which, having been reduced into fufion by fubterraneous heat, has been forcibly injected among the ftrata already confolidated, was fo different from that of other mineralogifts, that it feemed particularly to require farther examination. He concluded, that if this account was juft, fome confirmation of it muft appear at thofe places where the granite and the ftrata are in contact, or where the former emerges from beneath the latter. In fuch fituations, one might expect veins of the ftone which had been in fufion to penetrate into the ftone which had been folid; and fome imperfect defcriptions of granitic veins gave reafon to imagine that this phenomenon was actually to be obferved. Dr HUTTON was anxious that an *inftantia crucis* might fubject his theory to the fevereft teft.

ONE

latent, it will become colder than before. Thus alfo, when a quantity of heat afcends by any means whatever, from one ftratum of air to a fuperior ftratum, a part of it becomes latent, fo that an equilibrium of heat can never be eftablifhed among the ftrata; but thofe which are lefs, muft always remain colder than thofe that are more, compreffed. This was Dr HUTTON's explanation, and it contains no hypothetical principle whatfoever.

To one who confiders meteorology with attention, the want of an accurate hygrometer can never fail to be a fubject of regret. The way of fupplying this deficiency which Dr HUTTON practifed was by moiftening the ball of a thermometer, and obferving the degree of cold produced by the evaporation of the moifture. The degree of cold, *cæteris paribus*, will be proportional to the drynefs of the air, and affords, of courfe, a meafure of that drynefs. The fame contrivance, but without any communication whatfoever, occurred afterwards to Mr LESLIE, and being purfued through a feries of very accurate and curious experiments, has produced an inftrument which promifes to anfwer all the purpofes of photometry, as well as hygrometry, and fo to make a very important addition to our phyfical apparatus.

ONE of the places where he knew that a junction of the kind he wished to examine must be found, was the line where the great body of granite which runs from Aberdeen westward, forming the central chain of the Grampians, comes in contact with the schistus which composes the inferior ridges of the same mountains toward the south. The nearest and most accessible point of this line seemed likely to be situated not far to the eastward of Blair in Athol, and could hardly fail to be visible in the beds of some of the most northern streams which run into the Tay. Dr HUTTON having mentioned these circumstances to the Duke of ATHOL, was invited by that nobleman to accompany him in the shooting season into Glentilt, which he did accordingly, together with his friend Mr CLERK of Elden, in summer 1785.

THE Tilt is, according to the seasons, a small river, or an impetuous torrent, which runs through a glen of the same name, nearly south-west, and deeply intersects the southern ridges of the Grampian Mountains. The rock through which its bed is cut is in general a hard micaceous schistus; and the glen presents a scene of great boldness and asperity, often embellished, however, with the accompaniments of a softer landscape.

WHEN they had reached the Forest Lodge, about seven miles up the valley, Dr HUTTON already found himself in the midst of the objects which he wished to examine. In the bed of the river, many veins of red granite, (no less, indeed, than six large veins in the course of a mile), were seen traversing the black micaceous schistus, and producing, by the contrast of colour, an effect that might be striking even to an unskilful observer. The sight of objects which verified at once so many important conclusions in his system, filled him with delight; and as his feelings, on such occasions, were always strongly expressed, the guides who accompanied him were convinced that it must be nothing

less

lefs than the difcovery of a vein of filver or gold, that could call forth fuch ftrong marks of joy and exultation.

Dr HUTTON has defcribed the appearances at this fpot in the third volume of the Edinburgh *Tranfactions*, p. 79. and fome excellent drawings of them were made by Mr CLERK, whofe pencil is not lefs valuable in the fciences than in the arts. On the whole, it is certain, that of all the junctions of granite and fchiftus which are yet known, this at Glentilt fpeaks the moft unambiguous language, and moft clearly demonftrates the violence with which the granitic veins were injected among the fchiftus *.

In the year following, Dr HUTTON and Mr CLERK alfo vifited Galloway, in fearch of granitic veins, which they found at two different places, where the granite and fchiftus come in contact. One of thefe junctions was afterwards very carefully examined by Sir JAMES HALL and Mr DOUGLAS, now Lord SELKIRK, who made the entire circuit of a tract of granite country, which reaches from the banks of Loch Ken, where the junction is beft feen, weftward to the valley of Palnure, occupying a fpace of about 11 miles by 7. See *Edinburgh Tranf-actions*, vol. III. *Hiftory*, p. 8.

IN fummer 1787, Dr HUTTON vifited the ifland of Arran in the mouth of the Clyde, one of thofe fpots in which nature has collected, within a very fmall compafs, all the phenomena moft interefting to a geologift. A range of granite mountains, placed

ced

* I MUST take this opportunity of correcting a miftake which I have made in defcribing the junction in Glentilt, (*Illuftrations of the Huttonian Theory*, p. 310.) where I have faid, that the great body of granite from which thefe veins proceed, is not immediately vifible. This, however, is not the fact, for the mountains on the north fide of the glen are a mafs of granite to which the veins can be directly traced. This I have been affured of by Mr CLERK. Dr HUTTON has not defcribed it diftinctly; and not having feen the union of the veins with the granite on the north fide, when I vifited the fame fpot, I concluded too haftily, that it had not yet been difcovered.

ced in the northern part of the ifland, have their fides covered
with primitive fchiftus of various kinds, to which, on the fea-
fhore, fucceed fecondary ftrata of grit, limeftone, and even coal.
Here, therefore, Dr HUTTON had another opportunity of examining the junction of the granite and fchiftus, and found
abundance of the veins of the former penetrating into the latter.
In three different places he met with this phenomenon; in the
torrents that defcend from the fouth fide of Goatfield; in Glen-
rofa, on the weft, and in the little river Sannax, on the north-
eaft, of that mountain. From the firft of thefe he brought a
fpecimen of fome hundred weight, confifting of a block of
fchiftus, which includes a large vein of granite.

AT the northern extremity of the ifland he had likewife a
view of the fecondary ftrata lying upon the primary, with their
planes at right angles to one another. In the great quantity,
alfo, of pudding-ftone, containing rounded quartzy gravel, united by an arenaceous cement; in the multitude of whinftone
dikes, which abound in this ifland; and in the veins of pitch-
ftone, a foffil which he had not before met with in its native
place; he found other interefting fubjects of obfervation; fo
that he returned from this tour highly gratified, and ufed often
to fay that he had no where found his expectations fo much ex-
ceeded, as in the grand and inftructive appearances with which
nature has adorned this little ifland.

MR JOHN CLERK, the fon of his friend Mr CLERK of Elden,
accompanied him in this excurfion, and made feveral drawings,
which, together with a defcription of the ifland drawn up afterwards by Dr HUTTON, ftill remain in manufcript.

THE leaft complete of the obfervations at Arran was that
of the junction of the primitive with the fecondary ftrata,
which is but indiftinctly feen in that ifland, and only at
one place. Indeed, the contact of thefe two kinds of rock,
though it forms a line circumfcribing the bafes of all primitive
                                                    countries,

countries, is fo covered by the foil, as to be vifible in very few places. In the autumn of this fame year, however, Dr HUTTON had an opportunity of obferving another inftance of it in the bank of the river Jedd, about a mile above the town of Jedburgh. The fchiftus there is micaceous, in vertical plates, running from eaft to weft, though fomewhat undulated. Over thefe is extended a body of red fandftone, in beds nearly horizontal, having interpofed between it and the vertical ftrata a breccia full of fragments of thefe laft. Dr HUTTON has given an account of this fpot in the firft volume of his *Theory of the Earth*, p. 432., accompanied with a copper-plate, from a drawing by Mr CLERK.

IN 1788 he made fome other valuable obfervations of the fame kind. The ridge of the Lammer-muir Hills, in the fouth of Scotland, confifts of primary micaceous fchiftus, and extends from St Abb's-head weftward, till it join the metalliferous mountains about the fources of the Clyde. The fea-coaft affords a tranfverfe fection of this alpine tract at its eaftern extremity, and exhibits the change from the primary to the fecondary ftrata, both on the fouth and on the north. Dr HUTTON wifhed particularly to examine the latter of thefe, and on this occafion Sir JAMES HALL and I had the pleafure to accompany him. We failed in a boat from Dunglafs, on a day when the finenefs of the weather permitted us to keep clofe to the foot of the rocks which line the fhore in that quarter, directing our courfe fouthwards, in fearch of the termination of the fecondary ftrata. We made for a high rocky point or head-land, the SICCAR, near which, from our obfervations on fhore, we knew that the object we were in fearch of was likely to be difcovered. On landing at this point, we found that we actually trode on the primeval rock, which forms alternately the bafe and the fummit of the prefent land. It is here a micaceous fchiftus, in beds nearly vertical, highly indurated, and ftretch-

E

ing

ing from S. E. to N. W.   The furface of this rock runs with a
moderate afcent from the level of low-water, at which we land-
ed, nearly to that of high-water, where the fchiftus has a thin
covering of red horizontal fandftone laid over it ; and this fandftone, at the diftance of a few yards farther back, rifes into a
very high perpendicular cliff.   Here, therefore, the immediate
contact of the two rocks is not only vifible, but is curioufly dif-
fected and laid open by the action of the waves.   The rugged
tops of the fchiftus are feen penetrating into the horizontal beds
of fandftone, and the loweft of thefe laft form a breccia containing fragments of fchiftus, fome round and others angular, united by an arenaceous cement.

DR HUTTON was highly pleafed with appearances that fet in
fo clear a light the different formations of the parts which compofe the exterior cruft of the earth, and where all the circumftances were combined that could render the obfervation fatisfactory and precife.   On us who faw thefe phenomena for the
firft time, the impreffion made will not eafily be forgotten.   The
palpable evidence prefented to us, of one of the moft extraordinary and important facts in the natural hiftory of the earth,
gave a reality and fubftance to thofe theoretical fpeculations,
which, however probable, had never till now been directly au-
thenticated by the teftimony of the fenfes.   We often faid to
ourfelves, What clearer evidence could we have had of the different formation of thefe rocks, and of the long interval which
feparated their formation, had we actually feen them emer-
ging from the bofom of the deep ?   We felt ourfelves necef-
farily carried back to the time when the fchiftus on which we
ftood was yet at the bottom of the fea, and when the fandftone
before us was only beginning to be depofited, in the fhape of
fand or mud, from the waters of a fuperincumbent ocean.   An
epocha ftill more remote prefented itfelf, when even the moft
ancient of thefe rocks, inftead of ftanding upright in vertical
beds,

beds, lay in horizontal planes at the bottom of the sea, and was not yet disturbed by that immeasurable force which has burst asunder the solid pavement of the globe. Revolutions still more remote appeared in the distance of this extraordinary perspective *. The mind seemed to grow giddy by looking so far into the abyss of time; and while we listened with earnestness and admiration to the philosopher who was now unfolding to us the order and series of these wonderful events, we became sensible how much farther reason may sometimes go than imagination can venture to follow. As for the rest, we were truly fortunate in the course we had pursued in this excursion; a great number of other curious and important facts presented themselves, and we returned, having collected, in one day, more ample materials for future speculation, than have sometimes resulted from years of diligent and laborious research.

In the latter part of this same summer, (1788), Dr HUTTON accompanied the Duke of ATHOL to the Isle of Man, with a view of making a mineral survey of that island. What he saw there, however, was not much calculated to illustrate any of the great facts in geology. He found the main body of the island to consist of primitive schistus, much inclined, and more intersected with quartzy veins than the corresponding schistus in the south of Scotland. In two places on the opposite sides of the island, this schistus was covered by secondary strata, but the junction was no where visible. Some granite veins were observed in the schistus, and many loose blocks of that stone were met with in the soil, or on the surface, but no mass of it was to be seen in its native place. The direction of the primitive strata corresponded very well with that in Galloway, running near-

E 2

ly

ly from eaſt to weſt. This is all the general information which was obtained from an excurſion, which, in other reſpects, was very agreeable. Dr HUTTON performed it in company with his friend Mr CLERK, and they again experienced the politeneſs and hoſpitality of the ſame nobleman who had formerly entertained them, on an expedition which deſerves ſo well to be remembered in the annals of geology.

THOUGH from the account now given, it appears that Dr HUTTON's mind had been long turned with great earneſtneſs to the ſtudy of the theory of the earth, he had by no means confined his attention to that ſubject, but had directed it to the formation of a general ſyſtem, both of phyſics and metaphyſics *. He tells us himſelf, that he was led to the ſtudy of general phyſics, from thoſe views of the properties of body which had occurred to him in the proſecution of his chemical and mineralogical inquiries. In thoſe ſpeculations, therefore, that extended ſo far into the regions of abſtract ſcience, he began from chemiſtry ; and it was from thence that he took his departure in his circum-navigation both of the material and intellectual world.

THE chemiſt, indeed, is flattered more than any one elſe with the hopes of diſcovering in what the eſſence of matter conſiſts ; and Nature, while ſhe keeps the aſtronomer and the mechanician at a great diſtance, ſeems to admit him to more familiar converſe, and to a more intimate acquaintance with her ſecrets. The vaſt power which he has acquired over matter, the aſtoniſh-

ing

---

* AT what time theſe laſt ſpeculations began to ſhare his attention with the former, I have not been able to diſcover, though I have reaſon to believe that before I became acquainted with him, which was about 1781, he had completed a manuſcript treatiſe on each of them, the ſame nearly that he afterwards gave to the world. His ſpeculations on general phyſics were of a date much earlier than this.

THE Phyſical Syſtem, referred to here, forms the third part of a work, entitled, *Diſſertations on different Subjects in Natural Philoſophy,* in one vol. 4to, 1792.

ing transformations which he effects, his fuccefs in analyzing almoft all bodies, and in reproducing fo many, feem to promife
that he fhall one day difcover the effence of a fubftance which
he has fo thoroughly fubdued; that he fhall be able to bind
PROTEUS in his cave, and finally extort from him the fecret
of his birth; in a word, that he fhall find out what matter is,
of what elements it is compofed, and what are the properties effential to its exiftence.

IN entering upon this new inquiry, Dr HUTTON was forcibly
ftruck with the very juft reflection, That we do by no means explain the nature of body, when we defcribe it as made up of
fmall particles; becaufe if we allow to thefe particles any magnitude whatfoever, we do no more than affirm that great bodies
are made up of fmall ones. The elements of body muft, therefore, be admitted to be fomething unextended. To thefe unextended elements, Dr HUTTON gave the name of MATTER, and
carefully diftinguifhed between that term and the term BODY,
which he applied only to thofe combinations of matter that are
neceffarily conceived to poffefs impenetrability, extenfion and
inertia.

THE moft accurate examination of the properties of body con-
firms the truth of the opinion, that it is compofed of unextended
elements. Bodies may be compreffed into fmaller dimenfions;
many by the application of mechanical force, and all by the diminution of their heat: nor is there any limit to this compreffion, or
any point beyond which the farther reduction of volume becomes
impoffible. This holds of fubftances the moft compact, as well
as the moft volatile and elaftic, and clearly evinces that the elements of body are not in contact with one another, and that in
reality we perceive nothing in body but the exiftence of certain
powers or forces, acting with various intenfities, and in various
directions. Thus the fuppofed impenetrability, and of courfe
the extenfion of body, is nothing elfe than the effort of a refift-
ing

ing or repulfive power; its cohefion, weight, &c. the efforts of attractive power; and fo with refpect to all its other properties.

But if this be granted, and if it be true that in the material world every phenomenon can be explained by the exiftence of Power, the fuppofition of extended particles as a *fubftratum* or refidence for fuch power, is a mere hypothefis, without any countenance from the matter of fact. For if thefe folid particles are never in contact with one another, what part can they have in the production of natural appearances, or in what fenfe can they be called the refidence of a force which never acts at the point where they are prefent? Such particles, therefore, ought to be entirely difcarded from any theory that propofes to explain the phenomena of the material world.

Thus, it appears, that power is the effence of matter, and that none of our perceptions warrant us in confidering even body as involving any thing more than force, fubjected to various laws and modifications.

Matter, taken in this fenfe, is to be confidered as indefinitely extended, and without inertia. Its prefence through all fpace is proved by the univerfality of gravitation; and its want of inertia, by the want of refiftance to the planetary motions. Thus, in our inquiry concerning phyfical caufes, we are relieved from one great difficulty, that of fuppofing matter to act where it is not. The force of gravitation, according to this fyftem, is not the action of two diftant bodies upon one another, but it is the action of certain powers, diffufed through all fpace, which may be tranfmitted to any diftance. There feems to me, however, to remain a difficulty hardly lefs than that from which we appear to be relieved, *viz.* to affign a reafon why the intenfity with which fuch powers act on any body, fhould depend on the pofition and magnitude of all the bodies in the univerfe, and fhould bear to thefe continually the fame-relation. But, however this be, the ingenuity of Dr Hutton's reafonings cannot be queftioned, ed, nor, I think, the juſtneſs of many of his concluſions. His explanations of coheſion, heat, fluidity, deſerve particular attention. In one thing, however, he ſeems to have fallen into an error, which runs through much of his reaſoning, concerning the principles of gravitation and inertia. He affirms, that " without " gravity, a body endowed with all the other material qualities " would have no inertia ; that it would not diminiſh the velo- " city of the moving body by which it ſhould be actuated, nor " would it move a heavy body whatever were its velocity *." Now, this propoſition, though from its nature it cannot be brought to the immediate teſt of experience, is certainly inconſiſtent with the principles of mechanics ; at the ſame time, it is true, that we would not, in the caſe here ſuppoſed, have the ſame means of meaſuring the motion loſt, or gained by colliſion, which we have in the actual ſtate of bodies. This is perhaps what miſled Dr HUTTON ; and though his remarks on the meaſures of motion and force are very acute, and many of them very juſt, the mathematical reader will regret the want of that mode of reaſoning, which has raiſed mechanics to ſo high a rank among the ſciences.

IT is impoſſible not to remark the affinity of this theory with that of the celebrated BOSCOVICH, in which, as in this, all the phenomena of the material world are explained, by the ſuppoſition of forces variouſly modified, and without the aſſiſtance of ſolid or extended particles. Theſe forces are ſuppoſed to be arranged round mathematical points, which are moveable, and act on one another by means of the forces ſurrounding them. A moſt ingenious application of this principle is made to all the uſual reſearches of the mechanical philoſophy, and, it muſt be confeſſed, that few theories have more beauty and ſimplicity to recommend them, or do better aſſiſt the imagination in the ex-

planation

* *Diſſertations*, &c. p. 312. § 31.

planation of natural appearances. But it involves, in the whole
of it, this great difficulty, that mathematical points are not only
capable of motion, but capable of being endowed, or, at leaft,
diftinguifhed, by phyfical qualities. Dr HUTTON, in his theory, has avoided this difficulty, by giving no other than a negative definition of the MATTER which he fuppofes the elementa-
ry principle of body. On this account, though to the imagination his theory may want the charms which the other poffeffes,
yet it has the advantage of going juft to the extent to which our
perceptions or our obfervations authorife us to proceed, and of
being accurately circumfcribed by the limits pointed out by the
laws of philofophical induction *.

THE exiftence of matter neither heavy nor inert, which he
had taken fo much pains to eftablifh, was applied by him to explain

---

* THOUGH BOSCOVICH's Theory was publifhed long before Dr HUTTON's, fo
early, indeed, as the year 1758, there is no reafon to think that the latter was in
any degree fuggefted by the former. BOSCOVICH's theory was hardly known in
this country till about the year 1770, and the firft fketches of Dr HUTTON's theory
are of a much older date. Befides, the method of reafoning purfued by the authors
is quite different; and their conclufions, though alike in fome things, directly
contrary in others, as in what regards gravity, inertia, &c. The Monads of LEIB-
NITZ might more reafonably be fuppofed to have pointed out to Dr HUTTON the
neceffity of fuppofing the elements of body to be unextended, if the originality of
his own conceptions, and the little regard he paid to authority in matters of theory, did not relieve us from the neceffity of looking to others for the fources of his
opinions.

THE principal defect of his theory feems to me to confift in this, that it does not
ftate with precifion the difference between the conftitution of thofe *powers* which
fimply form matter, and thofe that form the more complex fubftance, body. In
other words, it does not explain what muft be added to matter to make it body.
The anfwer feems to me to be, that the addition of a repelling power, in all directions, is fufficient for that purpofe. Such a repulfion, if ftrong enough, would
produce both impenetrability and inertia. The matter, again, that poffeffed only
an attractive power, like *gravity*, or a repulfive power only in a certain direction, like *light*, would not be inert nor impenetrable. In this inference, however,
from his fyftem, I am not fure if I fhould meet with the author's approbation.

plain the phenomena of light, heat, and electricity. He considered all thefe three as modifications of the folar fubftance, and
thought that many of the appearances they exhibit, are only to
be explained on the fuppofition that they confift of an expanfive force, of which inertnefs is not predicable ; in particular,
that light is a power propagated from the fun in all directions,
like gravity, with this difference, that it is repulfive, while gravity is attractive, and requires time for its tranfmiffion, which
the latter does not, at leaft in any fenfible quantity *.

THE profecution of this fubject has led him to confider the nature
of PHLOGISTON, a fubftance once fo famous in chemiftry, but of
which the name has almoft as entirely difappeared from the vo-
cabulary of that fcience, as the word Vortex from the language
of phyfical aftronomy. The new and important experiments
made on the calcination of metals, and on the compofition of
water, are, as is well known, the foundations of the antiphlogiftic theory. Nobody was more pleafed than Dr HUTTON with
thefe experiments, nor held in higher eftimation the character
and abilities of the chemifts and philofophers by whom they
were conducted. He was neverthelefs of opinion, that the conclufions drawn from them are not altogether unexceptionable,
nor deduced with a fufficient attention to every circumftance.
This remark he thought peculiarly applicable to what regards
the compofition of water, to the phenomena of which experiment, the differtation we are now fpeaking of is chiefly directed.
The two aëriform fluids, it is there obferved, which compofe
water, in order to unite, muft not fimply be brought together,
for in that ftate they might remain for ever unchanged, but they
muft be fet on fire, and made to burn, and from this burning
there are evidently two fubftances which make their efcape,

F   namely,

* See Differtations V. and VI. on Matter and Motion, in the work above quot
ed. The Chemical Differtation on Phlogifton is in the fame volume, p. 171.

namely, Light and Heat. Though, therefore, the weight of the water generated, and of the gafes combined, may be admitted to be equal, yet it muft be acknowledged, that two fubftances are loft, which the chemift cannot confine in his clofeft veffel, nor weigh in his fineft balance, and it is going much farther than we are authorifed to do, either by experiment or analogy, to conclude that thefe fubftances have had no effect. As heat and light, in Dr HUTTON's fyftem, are compofed of that matter which does not gravitate, the exact coincidence which M. LA-VOISIER obferved between the weight of the water produced and that of the two elaftic fluids, united in the compofition of it, was no argument, in his eyes, againft the efcape of a very effen-tial part of the ingredients.

PURSUING the fame reafoning, he fhews how little ground there is to fuppofe that the heat and light evolved in this expe-riment proceed from the vital air; and he concludes, that the real explanation of the procefs is, that by burning, the matter of light and heat, or the phlogifton of the hydrogenous gas, is fet at liberty, and is thus enabled to unite with the vital air.

IN the fame manner, on examining what relates to the burn-ing of inflammable bodies, he finds the oxygenous gas unequal to the effect of furnifhing by its latent heat, or caloric, the whole of the fenfible heat that is produced. He concludes, therefore, that the hypothefis of the exiftence of phlogifton in thofe bodies that are termed inflammable, is neceffary to account for the pheno-mena of burning; phenomena, as he juftly remarks, which are among the moft curious and important of any that are exhibited by the material world. On the whole, it cannot be doubted, that great ingenuity and much found argument are difplayed throughout the whole of this differtation, and that whatever be ultimately decided with regard to the principle for which the author fo ftrenuoufly contends, he has made it evident, that the

conclufions

conclufions of the antiphlogiftic theory have been drawn with too much precipitancy, and carried farther than is warranted by the ftrict rules of inductive philofophy.

THE fubject of Fire, Light and Heat was refumed by Dr HUTTON feveral years after this period, and formed the fubject of a feries of papers which he read in the Royal Society of Edinburgh, and afterwards publifhed feparately. He there explains more fully his notion of the fubftances juft mentioned, which he confiders as different modifications of the folar matter, alike deftitute of inertnefs and of gravity.

A MORE voluminous work from Dr HUTTON's pen, made its appearance foon after the Phyfical Differtations, *viz. An Invefti-gation of the Principles of Knowledge, and of the Progrefs of Rea-fon from Senfe to Science and Philofophy*, in three volumes quarto.

HE informs us himfelf of the train of thought by which he was led to the metaphyfical fpeculations contained in thefe volumes. He had fatisfied himfelf, by his phyfical inveftigations, that body is not what it is conceived by us to be, a thing necef-farily poffeffing volume, figure and impenetrability, but merely an affemblage of powers, that by their action produce in us the ideas of thefe external qualities. His curiofity, therefore, was naturally excited to inquire farther into the manner in which we form our conceptions of body, or into the nature of the in-tercourfe which the mind holds with thofe things that exift without it. In purfuing this inquiry, he foon became convin-ced, that magnitude, figure and impenetrability, are no otherwife perceived by the mind than colour, tafte and fmell; that is, that what are called the primary qualities of body, are precifely on the fame footing with the fecondary, and are both conceptions of the mind, which can have no refemblance to the external caufe by which thofe conceptions are produced. The world, therefore, as conceived by us, is the creation of the mind itfelf, but of the mind acted on from without, and receiving informa-

tion

tion from fome external power.   But though, according to this
reafoning, there be no refemblance between the world without
us, and the notions that we form of it, though magnitude and
figure, though fpace, time and motion, have no exiftence but in
the mind; yet our perceptions being confiftent, and regu-
lated by conftant and uniform laws, are as much realities
to us, as if they were the exact copies of things really exifting;
they equally intereft our happinefs, and muft equally determine
our conduct.   They form a fyftem, not dependent on the mind
alone, but dependent on the action which certain external caufes
have upon it.   The whole doctrine, therefore, of moral obliga-
tion, remains the fame in this fyftem, and in that which main-
tains the perfect refemblance of our ideas to the caufes by which
they are produced.

MANY philofophers have regarded our ideas as very imperfect reprefentations of external things; but Dr HUTTON confiders their perfect diffimilitude as completely proved.   PLATO
has likened the mind to an eye, fo fituated, as to fee nothing
but the faint images of objects projected on the bottom of a dark
cave, while the objects themfelves are entirely concealed; but he
thinks, that by help of philofophy, the mental eye may be directed toward the mouth of the cave, and may perceive the objects in their true figure and dimenfions.   But, with Dr HUT-
TON, the figures feen at the bottom of the cave have no refemblance to the originals without; nor can man, by any contrivance, hold communication with thofe originals, nor ever know
any thing about them, except that they are not what they feem
to be, and have no property in common with the figures which
denote their exiftence.   In a word, external things are no more
like the perceptions they give rife to, than wine is fimilar to in-
toxication, or opium to the delirium which it produces.

IT has been already remarked, that this fyftem, however peculiar in other refpects, involves in it the fame principles of mo-

rals

rals with those more generally received ; and the same may be said as to the existence of God, and the immortality of the soul. The view which it presents of the latter doctrine, deserves particularly to be remarked. Death is not regarded here as the dissolution of a connexion between mind, and that system of material organs, by means of which it communicated with the external world, but merely as an effect of the mind's ceasing to perceive a particular order or class of things ; it is therefore only the termination of a certain mode of thought; and the extinction, not of any mental power, but of a train of conceptions, which, in consequence of external impulse, had existed in the mind. Thus, as nothing essential to intellectual power perishes, we are to consider death only as a passage from one condition of thought to another ; and hence this system appeared, to the author of it, to afford a stronger argument than any other, for the existence of the mind after death.

Indeed, Dr Hutton has taken great pains to deduce from his system, in a regular manner, the leading doctrines of morality and natural religion, having dedicated the third volume of his book almost wholly to that object. It is worthy of remark, that while he is thus employed, his style assumes a better tone, and a much greater degree of perspicuity, than it usually possesses. Many instances might be pointed out, where the warmth of his benevolent and moral feelings bursts through the clouds that so often veil from us the clearest ideas of his understanding. One, in particular, deserves notice, in which he treats of the importance of the female character to society, in a state of high civilization *. A felicity of expression, and a flow of natural eloquence, inspired by so interesting a subject, make us regret that his pen did not more frequently do justice to his thoughts.

The metaphysical theory, of which the outline, (though very imperfectly), has now been traced, cannot fail to recall the opinions

---

* *Investigation of the Principles of Knowledge,* Vol. III. p. 588. &c.

nions maintained by Dr BERKELEY concerning the exiftence of
matter. The two fyftems do indeed agree in one material point,
but differ effentially in the reft. They agree in maintaining,
that the conceptions of the mind are not copied from things of
the fame kind exifting without it ; but they differ in this, that Dr
BERKELEY imagined that there is nothing at all external, and
that it is by the direct agency of the Deity that fenfation and
perception are produced in the mind. Dr HUTTON holds, on
the other hand, that there is an external exiftence, from which
the mind receives its information, and by the action of which,
impreffions are made on it ; but impreffions that do not at all
refemble the powers by which they are caufed.

THE reafonings alfo by which the two theories are fupported,
are very diffimilar, though perhaps they fo far agree, that if Dr
BERKELEY had been better acquainted with phyfics, and had made
it more a rule to exclude all hypothefis, he would have arrived precifely at the fame conclufion with Dr HUTTON. Indeed, I cannot
help being of opinion, that every one will do fo, who, in invefti-
gating the origin of our perceptions, determines to reafon without affuming any hypothefis, and without taking for granted
any of thofe maxims which the mind is difpofed to receive, either, as fome philofophers fay, from habit, or, as others maintain, from an inftinctive determination, (fuch as has been termed *common fenfe*) that admits of no analyfis. Though this may
not be the kind of reafoning beft fuited to the fubject, yet it is fo
analogous to what fucceeds in other cafes, that it is good to have
an example of it, and, on that account, were it for nothing elfe, the
theory we are now fpeaking of certainly merits more attention
than it has yet met with *. The great fize of the book, and the
obfcurity

---

* I HAVE hardly found this work of Dr HUTTON's quoted by any writer of
eminence, except by Dr PAR, in his *Spital Sermon*, a tract no lefs remarkable for
learning and acutenefs, than for the liberality and candour of the fentiments which
it contains.

obfcurity which may juftly be objected to many parts of it, have probably prevented it from being received as it deferves, even among thofe who are converfant with abftract fpeculation. An abridgment of it, judicioufly executed, fo as to ftate the argument in a manner both perfpicuous and concife, would, I am perfuaded, make a valuable addition to metaphyfical fcience.

THE publication of this work was Dr HUTTON's occupation on his recovery from a fevere illnefs, with which he was feized in fummer 1793. Before this time he had enjoyed a long continuance of good health, and great activity both of body and mind. The diforder that now attacked him, (a retention of urine), was one of thofe that moft immediately threaten life, and he was preferved only by fubmitting to a dangerous, and painful operation. He was thus reduced to a ftate of great weaknefs, and was confined to his room for many months. By degrees, however, the goodnefs of his conftitution, aided, no doubt, by the vigour and elafticity of his mind, reftored him to a confiderable meafure of health, and rendered his recovery much more complete than could have been expected. One of his amufements, when he had regained fome tolerable degree of ftrength, was in fuperintending the publication, and correcting the proof-fheets of the work juft mentioned.

DURING his convalefcence, his activity was farther called into exertion, by an attack on his *Theory of the Earth*, made by Mr KIRWAN, in the *Memoirs of the Irifh Academy* *, and rendered formidable,

---

* THIS was not the firft attack which had been made on his theory, for M. DE LUC, in a feries of letters, inferted in the *Monthly Review* for 1790 and 1791, had combated feveral of the leading opinions contained in it. To thefe Dr HUTTON made no other reply, than is to be met with occafionally in the enlarged edition of his *Theory*, publifhed four years afterwards. If I do not miftake, however, he intended a more particular anfwer, and actually fent one to the editors

of

dable, not by the ftrength of the arguments it employed, but
by the name of the author, the heavy charges which it brought
forward, and the grofs mifconceptions in which it abound-
ed *.

BEFORE this period, though Dr HUTTON had been often urged by his friends to publifh his entire work on the *Theory of
the Earth*, he had continually put off the publication, and there
feemed to be fome danger that it would not take place in his
own life time. The very day, however, after Mr KIRWAN's
paper was put into his hands, he began the revifal of his manufcript, and refolved immediately to fend it to the prefs. The
reafon he gave was, that Mr KIRWAN had in fo many inftances
completely miftaken, both the facts, and the reafonings in his
Theory, that he faw the neceffity of laying before the world a
more ample explanation of them. The work was accordingly publifhed, in two volumes octavo, in 1795; and contained, befides what was formerly given in the *Edinburgh Tranfactions*, the proofs and reafonings much more in detail, and a
much fuller application of the principles to the explanation of
appearances. The two volumes, however, then publifhed, do
not complete the theory: a third, neceffary for that purpofe, remained behind, and is ftill in manufcript.

AFTER

of the fame *Review*, who refufed to infert it. This, indeed, I do not ftate with
perfect confidence, as I fpeak only from recollection, and would not, on that authority, bring a pofitive charge of partiality againft men who exercife a profeffion
in which impartiality is the firft requifite. Suppofing, however, the ftatement
here given to be correct, an excufe is ftill left for the Reviewers; they may fay,
that in communicating original papers, as they do not act in their judicial capacity, they are not bound to difpenfe juftice with their ufual blindnefs and feverity,
but may be permitted to relax a little from the exercife of a virtue that is fo often left to be its own reward.

* FOR a defence of Dr HUTTON againft the charges here alluded to, I muft
take the liberty of referring to the *Illuftrations of the Huttonian Theory*, p. 119.
and 125.

AFTER the publication of the work juſt mentioned, he began
to prepare another for the preſs, on a ſubjeƈt which had early
occupied his thoughts, and had been at no time of his life entirely negleƈted. This ſubjeƈt was huſbandry, on which he had
written a great deal, the fruit both of his reading and experience ;
and he now propoſed to reduce the whole into a ſyſtematic form,
under the title of *Elements of Agriculture*. This work, which he
nearly completed, remains in manuſcript. It is written with
conſiderable perſpicuity ; and though I can judge but very im-
perfeƈtly of its merits, I can venture to ſay, that it contains a
great deal of ſolid and praƈtical knowledge, without any of the
vague and unphiloſophic theory ſo common in books on the
ſame ſubjeƈt. In particular, I muſt obſerve, that where it treats
of climate, and the influence of heat, in accelerating the maturity
of plants, it furniſhes ſeveral views that appear to be perfeƈtly
new, and that are certainly highly intereſting.

THE period, however, was now not far diſtant, which was to
terminate the exertions of a mind of ſuch ſingular aƈtivity, and
of ſuch ardour in the purſuit of knowledge. Not long after the
time we are ſpeaking of, Dr HUTTON was again attacked by the
ſame diſorder from which he had already made ſo remarkable
a recovery. He was again ſaved from the danger that immediately threatened him, but his conſtitution had materially ſuffered,
and nothing could reſtore him to his former ſtrength. He
recovered, indeed, ſo far as to amuſe himſelf with ſtudy, and
with the converſation of his friends, and even to go on with the
work on agriculture, which was nearly completed. He was,
however, confined entirely to the houſe ; and in the courſe of
the winter 1796–7, he became gradually weaker, was extremely emaciated, and ſuffered much pain, but ſtill retained the
full aƈtivity and acuteneſs of his mind. He conſtantly employed himſelf in reading and writing, and was particularly pleaſed
with the third and fourth volumes of SAUSSURE's *Voyages*

aux Alpes, which reached him in the courfe of that winter, and became the laft ftudy of one eminent geologift, as they were the laft work of another. On Saturday the 26th of March he fuffered a good deal of pain; but, neverthelefs, employed himfelf in writing, and particularly in noting down his remarks on fome attempts which were then making towards a new mineralogical nomenclature. In the evening he was feized with a fhivering, and his uneafinefs continuing to increafe, he fent for his friend Mr RUSSEL, who attended him as his furgeon. Before he could poffibly arrive, all medical affiftance was in vain: Dr HUTTON had juft ftrength left to ftretch out his hand to him, and immediately expired.

Dr HUTTON poffeffed, in an eminent degree, the talents, the acquirements, and the temper, which entitle a man to the name of a philofopher. The direction of his ftudies, though in fome refpects irregular and uncommon, had been highly favourable to the developement of his natural powers, efpecially of that quick penetration, and that originality of thought, which ftrongly marked his intellectual character. From his firft outfet in fcience, he had purfued the track of experiment and obfervation, and it was not till after being long exercifed in this fchool, that he entered on the field of general and abftract fpeculation. He combined accordingly, through his whole life, the powers of an accurate obferver, and of a fagacious theorift, and was as cautious and patient in the former character, as he was bold and rapid in the latter.

LONG and continued practice had increafed his powers of obfervation to a high degree of perfection; fo that, in difcriminating mineral fubftances, and in feizing the affinities or dif-

ferences

ferences among geological appearances, he had an acutenefs hardly to be excelled. The eulogy fo happily conveyed in the Italian
phrafe, of *offervatore oculatiffimo*, might moſt juſtly be applied to
him ; for, with an accurate eye for perceiving the characters of
natural objects, he had in equal perfection the power of inter-
preting their fignification, and of decyphering thofe ancient hie-
roglyphics which record the revolutions of the globe. There
may have been other mineralogiſts, who could defcribe as well
the fracture, the figure, the fmell, or the colour of a fpecimen ;
but there have been few who equalled him in reading the characters, which tell not only what a foffil *is*, but what it *has been*,
and declare the feries of changes through which it has paffed.
His expertnefs in this art, the finenefs of his obfervations, and
the ingenuity of his reafonings, were truly admirable. It
would, I am perfuaded, be difficult to find in any of the fciences a better illuſtration of the profound maxims eſtabliſhed by
BACON, in his *Prærogativæ Inſtantiarum*, than were often afforded by Dr HUTTON's mineralogical difquifitions, when he exhibited his fpecimens, and difcourfed on them with his friends.
No one could better apply the luminous inſtances to elucidate
the obfcure, the decifive to interpret the doubtful, or the fimple
to unravel the complex. None was more ſkilful in marking the
gradations of nature, as ſhe paffes from one extreme to another ;
more diligent in obferving the *continuity* of her proceedings, or
more fagacious in tracing her footſteps, even where they were
moſt lightly impreffed.

WITH him, therefore, mineralogy was not a mere ſtudy of
names and external characters, (though he was fingularly well
verfed in that ſtudy alfo), but it was a fublime and important branch of phyfical fcience, which had for its object to un-
fold the connexion between the paſt, the prefent, and the future
conditions of the globe. Accordingly, his *collection of foffils* was
formed for explaining the principles of geology, and for illuſtra-

ting

ting the changes which mineral fubftances have gone through,
in the paffage which, according to all theories, they have made,
from a foft or fluid, to a hard and folid ftate, and from immer-
fion under the ocean, to elevation above its furface. The feries of thefe changes, and the relative antiquity of the different
fteps by which they have been effected, were the objects which
he had in view to explain; and his cabinet, though well adapted
to this end, with regard to other purpofes was very imperfect.
They who expect to find, in a collection, fpecimens of all the fpe-
cies, and all the varieties, into which a fyftem of artificial arrangement may have divided the foffil kingdom, will perhaps
turn faftidioufly from one that is not remarkable either for the
number or brilliancy of the objects contained in it. They, on
the other hand, will think it highly interefting, who wifh to reafon concerning the natural hiftory of minerals, and who are not
lefs eager to become acquainted with the laws that govern, than
with the individuals that compofe, the foffil kingdom.

THE lofs fuftained by the death of Dr HUTTON, was aggra-
vated, to thofe who knew him, by the confideration of how much
of his knowledge had perifhed with himfelf, and, notwithftanding
all that he had written, how much of the light collected by a
long life of experience and obfervation, was now completely ex-
tinguifhed. It is indeed melancholy to reflect, that with all who
make proficiency in the fciences founded on nice and delicate
obfervation, fomething of this fort muft unavoidably happen.
The experienced eye, the power of perceiving the minute differences, and fine analogies, which difcriminate or unite the objects
of fcience; and the readinefs of comparing new phenomena
with others already treafured up in the mind; thefe are accom-
plifhments which no rules can teach, and no precepts can put us
in poffeffion of. This is a portion of knowledge which every
man muft acquire for himfelf, and which nobody can leave as
an inheritance to his fucceffor. It feems, indeed, as if nature
had

had in this inftance admitted an exception to the rule, by which
fhe has ordained the perpetual accumulation of knowledge
among civilized men, and had deftined a confiderable portion
of fcience continually to grow up and perifh with the indivi-
dual.

A CIRCUMSTANCE which greatly diftinguifhed the intellectual character of the philofopher of whom we now fpeak, was
an uncommon activity and ardour of mind, upheld by the
greateft admiration of whatever in fcience was new, beautiful,
or fublime. The acquifitions of fortune, and the enjoyments
which moft directly addrefs the fenfes, do not call up more live-
ly expreffions of joy in other men, than hearing of a new invention, or being made acquainted with a new truth, would, at any
time, do in Dr HUTTON. This fenfibility to intellectual pleafure, was not confined to a few objects, nor to the fciences
which he particularly cultivated : he would rejoice over WATT's
improvements on the fteam-engine, or COOK's difcoveries in
the South Sea, with all the warmth of a man who was to fhare
in the honour or the profit about to accrue from them. The
fire of his expreffion, on fuch occafions, and the animation of
his countenance and manner, are not to be defcribed ; they were
always feen with great delight by thofe who could enter into his
fentiments, and often with great aftonifhment by thofe who
could not.

WITH this exquifite relifh for whatever is beautiful and fublime in fcience, we may eafily conceive what pleafure he derived from his own geological fpeculations. The novelty and
grandeur of the objects offered by them to the imagination,
the fimple and uniform order given to the whole natural hiftory
of the earth, and, above all, the views opened of the wifdom that
governs nature, are things to which hardly any man could be
infenfible ; but to him they were matter, not of tranfient delight,

but

but of folid and permanent happinefs. Few fyftems, indeed, were better calculated than his, to entertain their author with fuch noble and magnificent profpects; and no author was ever more difpofed to confider the enjoyment of them, as the full and adequate reward-of his labours.

THE great range which he had taken in fcience, has fufficiently appeared, from the account already given of his works *. There were indeed hardly any fciences, except the mathematical, to which he had not turned his attention, and his neglect of thefe probably arofe from this, that, at the time when his acquaintance with them fhould have commenced, his love of knowledge had already fixed itfelf on other objects. The aptitude of his mind for geometrical reafoning, was, however, proved on many occafions. His theory of rain refts on mathematical principles, and the conclufions deduced from them are perfectly accurate, though by no means obvious. I may add, that he had an uncommon facility in comprehending the nature of mechanical contrivances; and, for one who was not a practical engineer, could form, beforehand, a very found judgment concerning their effects.

NOTWITHSTANDING a tafte for fuch various information, and a mind of fuch conftant activity, he read but few fpeculative books, directing his attention chiefly to fuch as furnifhed the materials of fpeculation. Of voyages, travels, and books relating to

the

---

* He had ftudied with great care feveral fubjects of which no mention is made above. One of thefe was the Formation, or, as we may rather call it, the Natural Hiftory of Language. A portion of his metaphyfical work is dedicated to the Theory of Language, vol. I. p. 574, &c.; and vol. II. p. 624, &c. He read feveral very ingenious papers on the Written Language, in the Royal Society of Edinburgh, fee *Tranfactions of the Royal Society of Edinburgh*, vol. II. *Hift.* p. 5. &c. The Chinefe language, as an extreme cafe in the invention of writing, had greatly occupied his thoughts, and is the fubject of feveral of his manufcripts.

the natural hiſtory of the earth, he had an extenſive knowledge : he had ſtudied them with that critical diſcuſſion which ſuch books require above all others; carefully collecting from them the facts that appeared accurate, and correcting the narratives that were imperfect, either by a compariſon with one another, or by applying to them the ſtandard of probability which his own ob- ſervation and judgment had furniſhed him with. On the other hand, he beſtowed but little attention on books of opinion and the- ory ; and while he truſted to the efforts of his mind for digeſting the facts he had obtained from reading or experience, into a ſyſtem of his own, he was not very anxious, at leaſt till that was accom- pliſhed, to be informed of the views which other philoſophers had taken of the ſame ſubject. He was but little diſpoſed to con- cede any thing to mere authority ; and to his indifference about the opinions of former theoriſts, it is probable that his own ſpeculations owed ſome part, both of their excellencies, and their defects.

As he was indefatigable in ſtudy, and was in the habit of uſing his pen continually as an inſtrument of thought, he wrote a great deal, and has left behind him an incredible quantity of manuſcript, though imperfect, and never intended for the preſs. Indeed his manner of life, at leaſt after he left off the occupations of huſbandry, gave him ſuch a command of his time, as is en- joyed by very few. Though he uſed to riſe late, he began imme- diately to ſtudy, and generally continued buſy till dinner. He di- ned early, almoſt always at home, and paſſed very little time at table ; for he ate ſparingly, and drank no wine. After dinner he reſumed his ſtudies, or, if the weather was fine, walked for two or three hours, when he could not be ſaid to give up ſtudy, though he might, perhaps, change the object of it. The evening he al- ways ſpent in the ſociety of his friends. No profeſſional, and rarely any domeſtic arrangements interrupted this uniform courſe of life, ſo that his time was wholly divided between the purſuits

of

of fcience and the converfation of his friends, unlefs when he travelled from home on fome excurfion, from which he never failed to return furnifhed with new materials for geological inveftigation.

To his friends his converfation was ineftimable; as great talents, the moft perfect candour, and the utmoft fimplicity of character and manners, all united to ftamp a value upon it. He had, indeed, that genuine fimplicity, originating in the abfence of all felfifhnefs and vanity, by which a man lofes fight of himfelf altogether, and neither conceals what is, nor affects what is not. This fimplicity pervaded his whole conduct; while his manner, which was peculiar, but highly pleafing, difplayed a degree of vivacity, hardly ever to be found among men of profound and abftract fpeculation. His great livelinefs, added to this aptnefs to lofe fight of himfelf, would fometimes lead him into little excentricities, that formed an amufing contraft with the graver habits of a philofophic life.

Though extreme fimplicity of manner does not unfrequently impart a degree of feeblenefs to the expreffion of thought, the contrary was true of Dr Hutton. His converfation was extremely animated and forcible, and, whether ferious or gay, full of ingenious and original obfervation. Great information, and an excellent memory, fupplied an inexhauftible fund of illuftration, always happily introduced, and in which, when the fubject admitted of it, the witty and the ludicrous never failed to occupy a confiderable place.—But it is impoffible by words to convey any idea of the effect of his converfation, and of the impreffion made by fo much philofophy, gaiety and humour, accompanied by a manner at once fo animated and fo fimple. Things are made known only by comparifon, and that which is *unique* admits of no defcription.

The whole exterior of Dr Hutton was calculated to heighten the effect which his converfation produced. His figure was

flender

flender, but indicated activity; while a thin countenance, a high forehead, and a nofe fomewhat aquiline, befpoke extraordinary acutenefs and vigour of mind. His eye was penetrating and keen, but full of gentlenefs and benignity; and even his drefs, plain, and all of one colour, was in perfect harmony with the reft of the picture, and feemed to give a fuller *relief* to its characteriftic features *.

THE friendfhip that fubfifted between him and Dr BLACK has been already mentioned, and was indeed a diftinguifhing circumftance in the life and character of both. There was in thefe two excellent men that fimilarity of difpofition which muft be the foundation of all friendfhip, and, at the fame time, that degree of diverfity, which feems neceffary to give to friends the higheft relifh for the fociety of one another.

THEY both cultivated nearly the fame branches of phyfics, and entertained concerning them nearly the fame opinions. They were both formed with a tafte for what is beautiful and great in fcience; with minds inventive, and fertile in new combinations. Both poffeffed manners of the moft genuine fimplicity, and in every action difcovered the fincerity and candour of their difpofitions; yet they were in many things extremely diffimilar. Ardour, and even enthufiafm, in the purfuit of fcience, great rapidity of thought, and much animation, diftinguifhed Dr HUTTON on all occafions. Great caution in his reafonings, and a coolnefs of head that even approached to indifference, were characteriftic of Dr BLACK. On attending to their converfation, and the way in which they treated any queftion of fcience or philofophy, one would fay that Dr BLACK dreaded nothing fo much as error, and that Dr HUTTON dread-

H ed

* A PORTRAIT of Dr HUTTON, by RAEBURN, painted for the late JOHN DA-VIDSON, Efq; of Stewartfield, one of his old and intimate friends, con eys a good idea of a phyfiognomy and character of face to which it was difficult to do complete juftice.

ed nothing ſo much as ignorance; that the one was always afraid of going beyond the truth, and the other of not reaching it. The curioſity of the latter was by much the moſt eaſily awakened, and its impulſe moſt powerful and imperious. With the former, it was a deſire which he could ſuſpend and lay aſleep for a time; with the other, it was an appetite that might be ſatisfied for a moment, but was ſure to be quickly renewed. Even the ſimplicity of manner which was poſſeſſed by both theſe philoſophers, was by no means preciſely the ſame. That of Dr BLACK was correct, reſpecting at all times the prejudices and faſhions of the world; that of Dr HUTTON was more careleſs, and was often found in direct colliſion with both.

FROM theſe diverſities, their ſociety was infinitely pleaſing, both to themſelves and thoſe about them. Each had ſomething to give which the other was in want of. Dr BLACK derived great amuſement from the vivacity of his friend, the ſallies of his wit, the glow and original turn of his expreſſion; and that calmneſs and ſerenity of mind which, even in a man of genius, may border on languor or monotony, received a pleaſing impulſe by ſympathy with more powerful emotions.

ON the other hand, the coolneſs of Dr BLACK, the judiciouſneſs and ſolidity of his reflections, ſerved to temper the zeal, and reſtrain the impetuoſity of Dr HUTTON. In every material point of philoſophy they perfectly agreed. The theory of the earth had been a ſubject of diſcuſſion with them for many years, and Dr BLACK ſubſcribed entirely to the ſyſtem of his friend. In ſcience, nothing certainly is due to authority, except a careful examination of the opinions which it ſupports. It is not meant to claim any more than this in favour of the Huttonian Geology; but they who reject that ſyſtem, without examination, would do well to conſider, that it had the entire and unqualified approbation of one of the cooleſt, and

ſoundeſt

foundeſt reaſoners of which the preſent age furniſhes any example.

Mr CLERK of Elden was another friend, with whom, in the formation of his theory, Dr HUTTON maintained a conſtant communication. Mr CLERK, perhaps from the extenſive property which his family had in the coal-mines near Edinburgh, was early intereſted in the purſuits of mineralogy. His inquiries, however, were never confined to the objects which mere ſituation might point out, and, through his whole life, have been much more directed by the irreſiſtible impulſe of genius, than by the action of external circumſtances. Though not bred to the ſea, he is well known to have ſtudied the principles of naval war with unexampled ſucceſs; and though not exerciſing the profeſſion of arms, he has viewed every country through which he has paſſed with the eye of a ſoldier as well as a geologiſt. The intereſt he took in ſtudying the ſurface no leſs than the interior of the earth; his extenſive information in moſt branches of natural hiſtory; a mind of great reſource, and great readineſs of invention; made him, to Dr HUTTON, an invaluable friend and co-adjutor. It cannot be doubted, that, in many parts, the ſyſtem of the latter has had great obligations to the ingenuity of the former, though the unreſerved intercourſe of friendſhip, and the adjuſtments produced by mutual ſuggeſtion, might render thoſe parts undiſtinguiſhable even by the authors themſelves. Mr CLERK's pencil was ever at the command of his friend, and has certainly rendered him moſt eſſential ſervice.

BUT it was not to philoſophers and men of ſcience only that Dr HUTTON's converſation was agreeable. He was little known, indeed, in general company, and had no great reliſh for the enjoyment which it affords; yet he was fond of domeſtic ſociety, and took great delight in a few private circles, where ſeveral excellent and accompliſhed individuals of both ſexes thought themſelves happy to be reckoned in the number

ber of his friends. In one or other of thefe, he was accuftomed almoft every evening to feek relaxation from the ftudies of the day, and found always the moft cordial welcome. A brighter tint of gaiety and chearfulnefs fpread itfelf over every countenance when the Doctor entered the room; and the philofopher who had juft defcended from the fublimeft fpeculations of metaphyfics, or rifen from the deepeft refearches of geology, feated himfelf at the tea-table, as much difengaged from thought, as chearful and gay, as the youngeft of the company. Thefe parties were delightful, and, by all who have had the happinefs to be prefent at them, will never ceafe to be remembered with pleafure.

He ufed alfo regularly to unbend himfelf with a few friends, in the little fociety alluded to in Profeffor STEWART's *Life of* Mr SMITH, and ufually known by the name of the *Oyfter Club*. This club met weekly; the original members of it were Mr SMITH, Dr BLACK, and Dr HUTTON, and round them was foon formed a knot of thofe who knew how to value the familiar and focial converfe of thefe illuftrious men. As all the three poffeffed great talents, enlarged views, and extenfive information, without any of the ftatelinefs and formality which men of letters think it fometimes neceffary to affect; as they were all three eafily amufed; were equally prepared to fpeak and to liften; and as the fincerity of their friendfhip had never been darkened by the leaft fhade of envy; it would be hard to find an example, where every thing favourable to good fociety was more perfectly united, and every thing adverfe more entirely excluded. The converfation was always free, often fcientific, but never didactic or difputatious; and as this club was much the refort of the ftrangers who vifited Edinburgh, from any object connected with art or with fcience, it derived from thence an extraordinary degree of variety and intereft. It

is

is matter of real regret, that it has been unable to furvive its founders.

The fimplicity of manner that has been already remarked as fo ftrikingly exemplified in Dr Hutton, was but a part of an extreme difintereftednefs which manifefted itfelf in every thing he did. He was upright, candid, and fincere ; ftrongly attached to his friends ; ready to facrifice any thing to affift them ; humane and charitable. He fet no great value on money, or, perhaps, to fpeak properly, he fet on it no more than its true value ; yet, owing to the moderation of his manner of life, and the ability with which his friend Mr Davie conducted their joint concerns, he acquired confiderable wealth.

He was never married, but lived with his fifters, three excellent women, who managed his domeftic affairs ; and of whom, only one, Mifs Isabella Hutton, remained to lament his death. By her his collection of foffils, about which he left no particular inftructions, was prefented to Dr Black ; who thought that he could not better confult the advantage of the public, or the credit of his friend, than by giving it to the Royal Society of Edinburgh, under the condition that it fhould be completely arranged, and kept for ever feparate, for the purpofe of illuftrating the Huttonian Theory of the Earth.

FINIS.

Printed in the United States
By Bookmasters